ENCYCLOPAEDIA OF HORTICULTURAL CROPS

Volume-2

HORTICULTURAL CROP PRODUCTION

S. K. GUPTA

RAJAT PUBLICATIONS
NEW DELHI - 110 002 (INDIA)

RAJAT PUBLICATIONS
4740/23, Ansari Road, Daryaganj
New Delhi- 110 002 (India)
Phones : +91-11-23267924, +91-9811484973
E-mail: rajat_publication@yahoo.com

Encyclopaedia of Horticultural Crops

© *Reserved*

Edition 2025

ISBN 978-81-7880-319-7 (Set)

[The responsibility for facts stated opinion expressed or conclusions reached and plagiarism, if any, in this volume is entirely that of the Editor. The publisher bears no responsibility for them whatsoever.]

PRINTED IN INDIA

Published by Mrs. Seema Wasan for Rajat Publications, New Delhi and Printed at H.S. Offset Press, Delhi.

Preface

Horticulture refers to the area of agricultural science that deals with garden plant cultivation, the science and out of growing fruits, vegetables, ornamental trees bushes and flowers, on a commercial scale, in nurseries, orchards and gardens. Along with forestry and agronomy, horticulture constitutes the three broad branches of agricultural science and practice. As importance and relevance of aggressive cannot be ignored, so can that of horticulture be also not ignored. In fact, with the increase in demand for newer and better varieties of horticultural products, the significance of horticulture has increased all the more.

This encyclopedia of horticultural crops is a three volume set which brings a compendious compilation of information on the various forms science and practice of cultivating horticultural crops. Through these three works, foundation has been laid by the encyclopedia for a discussion of the principles of horticultural science, what it comprises of, its history and its present, and what the future holds for this field.

The encyclopedia commences with the book organic vegetable production a book which seeks to discuss the cultivation of organic vegetables in agro-science. As organic farming and its products have entered the market in a significant manner, it becomes imperative that a manual seek the principles of this practice and therein lies the purpose of this book. Cultivation of Herbs and Flowers as the title suggests, is a work bringing forth a comprehensive detailing of cultivating herbs and flower which after vegetables, consultant a major part of the demand for horticultural products, both for purposes in practical applications (as in medicinal), and for ornamentation. The concluding work in this series Is the book Horticultural Crop Production. As many horticultural crops today particularly vegetables—are often grown as field crops and considered agricultural commodities, it has become increasingly evident that horticulture products are much more importance than as mere garden products. This book seeks a study of horticultural science as a whole, with special emphasis on its products as "Crops".

All these three books aim at providing a reliable encylopaedia on horticultural productions; and provide material for through study of the science itself. It is hoped that the comprehensive and informative content of this series foster an enriching reading and learning experience for readers.

S.K. Gupta

Contents

1
Crop Production Technologies

The intensification of crop production improved the level of food consumption in most developing countries during the last 30 years despite a world population growth of 2.4 billion in the same period. However, the intensification also led to a number of disadvantageous side-effects like erosion, deterioration of soil quality, salinisation, and others.

Every year, about 1 % of the cultured area has to be replaced because of these reasons. Especially in developed countries, the use of agrochemicals is a growing public concern. The amount of crop protection products used has risen steadily during the past 30 years. Presently, more than 25 billion US$ per year are spent for herbicides, fungicides, insecticides, soil disinfectants and growth regulators. In Western Europe, the most crop protection products are used per area of arable land, followed by the USA.

Plant genetic engineering may contribute substantially to the development of new pest and disease resistant varieties by increasing the speed of conventional breeding and by transferring useful genes from various plants and other organisms into crops.

In Asia, most vegetable growers have a land holding which is smaller than one hectare (2.4 acres). They are interested in any technology which helps them to produce more crops each year from their land, particularly during the off season when prices are higher. For this reason, greenhouses and other types of structures have become popular among Asian growers. Production under structures not only increases the total annual yield per unit area, but also improves the quality. Vegetables have a better flavor and appearance when they are grown in a controlled environment, where growers can make precise adjustments of nutrients and other inputs.

The kinds of protective structures used by growers in Asia range from simple structures such as rain shelters, shadehouses, mulches, row covers and plastic tunnels, to permanent structures covered in plastic or glass with computerised environmental controls. Growers in countries where typhoons are common over the summer tend to prefer low-cost structures which can be quickly and cheaply replaced. Greenhouses in such countries generally have walls and roof of plastic rather than glass, over a metal or even a bamboo framework.

The discovery and development of the polyethylene polymer in the late 1930s, and its subsequent introduction in the early 1950s in the form of plastic films, mulches, and drip-irrigation tubing and tape, revolutionised the commercial production of many vegetable crops. The later discovery of other polymers, such as polyvinyl chloride, polypropylene, and polyesters, and their use in pipes, fertigation equipment, filters, fittings and connectors, and row covers, further extended the use of plastic in agriculture.

In general, structures are used to overcome low temperatures in temperate Asian countries, and high temperatures in tropical ones. Modifying the temperature is required, not only for crop protection, but also to provide growers with more comfortable working conditions. A common use of structures in both cool and hot climates is for the nursery production of high-quality and/or early seedlings for transplanting.

In Japan, greenhouses were first introduced to protect crops from cold weather. Now greenhouses are commonly used all year round. High summer temperatures are modified by good ventilation, shading and evaporative cooling. New covering materials are being developed which block solar radiation. In Malaysia, the area used for the protected cultivation of horticultural crops has expanded rapidly over the last decade.

The most important constraint is marketing. In general, vegetables produced under structures cannot successfully compete with those produced in the open field imported from neighboring countries, unless stringent measures are taken by the authorities to safeguard the quality and safety of vegetables. These include specifying the maximum levels of pesticide residues, and accurate labeling.

Greenhouse production of vegetables in Korea has made much progress in terms of the external design and the construction of modernised glasshouses. However, there is still a need to reduce the labor cost of operations such as transplanting and harvesting. Protected cultivation of vegetables takes 2-3 times more labor than open field cultivation. Mechanisation of some operations may be necessary. This in turn implies standard facilities where machines can be used.

In Taiwan, some fruits such as peaches are being produced under structures. Establishing a permanent orchard structure is expensive, and needs a considerable amount of technical skill. The system must be able to meet the environmental requirements of the fruit grown, so that trees produce a stable yield of high-quality produce.

In Thailand, protective structures are being used to produce flowers by the hill tribes of northern Thailand, under the Royal Project Foundation. At present, the Royal Project Foundation has a large planting area under plastic in Thailand, 90% of which is used for flower production. Greenhouse designs developed by the Foundation have already been widely used nationwide. Structures for some fruit crops such as grape and peach are under investigation and show promising results, while research on vegetables under structures is also under way.

While greenhouse production increases the net income of farmers in Northern Thailand, they still need a great deal of outside technical support. Greenhouse production is so different from their traditional agriculture, that it will take some time for growers to acquire all the needed technical skills. This is a common problem elsewhere in Asia. Production under structures requires good management. Proper planning and attention to detail are needed, if the grower is to succeed

Another widespread problem is that small-scale farmers may find it difficult to finance the investment cost of even a simple structure. In the Philippines, for example, the initial cost of establishing protective structures such as greenhouses and plastic houses is considerable, and not within the reach of ordinary farmers. Designs using cheaper materials are needed. Furthermore, mulching materials which are durable and biodegradable must be developed to protect the environment. Further studies are also needed on protective structures which can withstand typhoons, and which have a better ventilation system for a tropical climate.

GREENHOUSE MECHINERY

Whereas many Asian smallholders have adopted machinery for rice production, machines are not much used under structures. There are several reasons for this. Firstly, most farms are too small. Even if suitable machines were available, the utilisation rate would be very low. Secondly, growers produce a wide range of crops. It is not easy to develop machines that can take care of a wide range of needs of a variety of crops. Thirdly, most greenhouses are temporary structures with narrow walkways which are difficult for machines to enter.

However, the scale of farming in most Asian countries is expected to increase, particularly in highly industrialised ones. Many growers are enlarging the size of their operations through group farming or producers' associations. Designs for greenhouse machinery should focus on attachments which can be adapted to various crops and different operations, and simple, lightweight machinery suitable for a rural labor force which includes many women and old people. Eventually, new cultivation models with uniform beds and standardised seedling trays can be expected for mechanised operations.

In vegetable production in greenhouses, continuous cultivation leads to heavy losses, particularly from soil-borne diseases and nematodes. Grafted seedlings have become popular in Japan. Grafting gives greater disease tolerance and increased vigor to crops, to an extent which justifies the higher cost of such seedlings. In 1990, in areas producing watermelons, cucumbers, melons, tomatoes and eggplants, 59% of plants were grafted. A range of rootstock species and varieties are used, and various grafting methods. Recently, tube grafting has become popular in the manual grafting of tomato, eggplant and cucumber plants, because it is faster. Grafting robots and healing chambers have been developed for use in plant nurseries.

Plug Culture

In plug culture, vegetables are sown by direct seeding into small cells or plugs. Usually a single seed is planted in each plug. When the seedlings have grown to the 2-3 leaf stage, they are transplanted into flats. These may contain from 18 to more than a hundred cells full of growing medium. The flats are usually left to grow under netting or some kind of structure. The plug system is widely used in Taiwan for the production of leafy vegetables, which are harvested at an early stage when they are still young and tender.

Consumers like the flavor of the vegetables, and the fact that they are free of field soil. From the grower's point of view, plug seedlings have a faster growth rate, a shorter cropping period and a longer storage life. Research data also shows that vegetables grown in a porous medium have better root and shoot growth than those grown in field soil, resulting in higher yields. However, the system is labor intensive, particularly transplanting the plugs into the flats and taking them to the field. Future research should focus on mechanisation and automation of the system to reduce the labor demand. Another useful innovation would be biodegradable or photodegradable flats that are more environmentally friendly. The possibility of using colored flats which reflect light, thus promoting plant growth and repelling insects, should also be explored.

Plug production technology was introduced into Korea in 1992, and is now an important industry. The technology was first used for the mass production of vegetable seedlings to be transplanted into greenhouses and the open field. This gave growers a steady supply of high-quality seedlings, and relieved them of the burden of having to produce their own planting materials. Recently, there has been growing interest in flower plug transplants.

Soilless System

Since the early 1980s, many soilless growing systems have been developed in Asia, some suited to tropical countries and some to temperate ones. Some of these were based on hydroponics. Early types were open systems. More recent ones are closed. Drained water is collected and brought to a central poinmt for reuse, after the water has been disinfected and its EC and pH adjusted. Nowadays, however, 80% of greenhouse producers are using some kind of artificial medium for their crops. Soilless systems using media of various kinds systems give growers good control over plant nutrients, so that produce has good appearance, flavor and nutritive value. Porous media also offer an ideal environment for root growth.

Agricutlural Technology

Agricultural production is a prerequisite to rural poverty alleviation. However, in itself, agricultural growth is not sufficient to eradicate poverty, nor can it create adequate opportunities for enhanced rural employment, income generation, and economic access to food.

Agro-processing maintains a high potential for boosting economies based on agriculture to help ensure sustainable growth and increasing rural incomes in the region. And while the new global economy ushers in a rapid expansion of the worldwide distribution of agricultural products, our concern now is not only to produce food in sufficient quantity, but also to enhance the competitiveness of the Asian agricultural industry in the world market by developing value-added foods based on higher quality standards and improved processing techniques.

In the last half of the 20th century, rapid industrialisation and urbanisation have been consuming too much natural energy and resulting in rapid migration of humans and the fast movement of resources and information. Agriculture is not exempted from this fast change and it is now shifting from small-scale to commercial and even to global production. Thus, new agricultural technologies have been developed in order to supply stable and sufficient food to the expanding population.

Meanwhile, sustainable agriculture as well as the preservation of the diverse food culture and maximum utilisation of food and biomass must be ensured. East Asia in the Pacific Rim is the origin of diverse agricultural technology and food culture, and is envisioned to be at the forefront of scientific and technological advancement in agricultural production.

The extensive application of diverse agricultural technologies is vital in order to meet the demands of an increasing population. Agricultural production in the Asian region is notably decreasing, with the rate of increase in food production less than that of the increase in population, particularly in China and India. Unless changes in agricultural policies are made, the threat of hunger and famine will continue to loom in some developing countries in the region.

More efficient land use is necessary in view of the increasing demand in agricultural products and the decreasing rate in annual production. Optimum production rate must be met in order to keep the sustainable use of land, i.e. crop yield and fertiliser use must be balanced.

Increased crop yield must also be achieved through the use of better varieties with higher photosynthetic activities, and better utilisation of phosphorus, potassium, and other minerals under arid conditions such as low or high pH soil, high salinity, and undesirable water conditions. These varieties must be developed as soon as possible in order to cope with the rapid increase in population and erosion of fertile soils through new biotechnology such as recombinant technology and land management.

An understan-ding of agroecosystems is the key to determining effective farming systems. A report of results from a 21-year study of agronomic and ecological performance of biodynamic, bioorganic, and conventional farming systems in Central Europe showed crop yields to be 20% lower in the organic systems, although input of fertiliser and energy was reduced by 34% to 53% and pesticide input by 97%. Enhanced soil fertility and higher biodiversity found in organic plots may render these systems less dependent on external inputs.

Improvement of yields of crops can also be achieved by optimising traditional or conventional farming, organic farming, and bio-farming or the use of genetically modified crops. These three farming types can co-exist depending on the farming and land conditions. They can complement each other and actually are not competitive, for each will have its own market. It is obvious however that organic farming will cost much in terms of label and management so that the selling price is naturally higher. Traditional or conventional farming may sacrifice sustainability of land use. Thus, bio-farming and good management is the one expected to give the least cost and best sustainability. However, recombinant crops must first be proven to be safe to humans, animal and the environment.

Traditional foods are a treasure box of resources in terms of material and knowledge. The way they are produced, the local substrates used in the countries where they are developed, and the presence of diverse microorganisms in these foods are good and novel resources for the development of technology for the food industry. These microbes and fermented products have various functionalities in maintaining human health, thus, have great potential in the development of physiologically functional foods. Oligosaccharides with mineral enhancing capability, prebiotic activity, anti-constipation, peptides with anti-hypertensive, improvements of lipid metabolisms, anti-obesity, and cholesterol-lowering activity are found in various Asian traditional foods.

Biotechnology in Production

Over the past two decades, the technology for producing seeds and seedlings has developed very rapidly. Many new techniques have been incorporated into production systems, and have benefited farmers all over the world. However, this rapid pace of technical development means that scientists and technicians involved in the production of high-quality seeds are constantly having to update their skills.

Genetic diversity is the basis of plant breeding. Precious biodiversity is being lost with the disappearance of numerous wild plant species, mainly because of overexploitation and loss of habitat. Many wild plants have medicinal uses. They tend to become more valuable the scarcer they become. This encourages people to gather them to sell, and puts heavy pressure on surviving populations.

Tissue culture techniques are being widely used for the propagation and conservation of many traditional Chinese medical plants. Some of them are becoming rare in the wild. The conservation of germplasm in vitro keeps these species safe from accidental extinction. The techniques also make available a large body of material for pharmacological studies and chemical analysis. They also provide large quantities of disease-free seedlings for farmers. Tissue culture has made possible the mass production of disease-free and uniform plants. The technique thus brings farmers the great benefit of high-quality planting materials of new high-value crops

Finally, these same techniques can be used for the safe transfer of germplasm from one country to another. Many of the diseases and pests which damage agriculture today were introduced accidentally when new crop varieties were imported from overseas. So much damage has already been done from these accidental invasions, that it is becoming difficult for countries to exchange plant materials. Many governments tend to feel that the possible damage from introductions may well outweigh any potential benefits. However, plant breeding programs need access to new germplasm. Tissue culture offers a solution to the problem. Germplasm can cross international boundaries safely, in the assurance that no pests and diseases are secretly traveling with it.

In recent years, remarkable progress in molecular biology techniques has enabled scientists to acquire a more detailed understanding of plant pathogens. This training course introduced participants to current techniques of serological and molecular diagnostic techniques.

Serological assay is based on antibodies produced by animals, usually rabbit or rats. When suitable antibodies are available, the most sensitive and efficient test for plant virus is ELISA (enzyme-linked immunosorbent assay). ELISA is widely used, since it is easy to apply and does not need any expensive equipment.

Another testing method is PCR (polymerase chain reaction). PCR testing is very accurate, but it is also rather expensive and technically difficult. The participants were shown the preparation of samples for PCR testing, including the precautions needed to avoid the contamination of samples. They then practiced the steps themselves in laboratory exercises.

The DNA composition of an individual plant is what determines its essential characteristics. The DNA fingerprint describes the DNA sequence profile of an individual. This is the same in every cell of that individual. The DNA fingerprint cannot be altered by any change in the environment. Nor can it be changed by anything that happens to an individual during its lifetime.

Different DNA markers are available, including RFLP (restriction fragment length polymorphism), RAPD (random amplified polymorphic DNA) and AFLP (amplified fragment length polymorphism). These markers are used to characterise a variety or line, and in this way protect the intellectual property rights of the plant breeder. They are also used to evaluate the purity of seeds, and to detect genetic drift in germplasm collections. It is also advisable to use the material for only five generations, or only five sub-culturing cycles.

Improving Water Resources

Infiltration depends on there being sufficient porosity in

the surface soil for rainfall to infiltrate, and in the subsoil and parent material for rainwater to percolate. When the porosity of the surface soil is too low to accept rainfall, or subsoil porosity is too low to allow rainwater percolation (i.e. permeability is too slow), then infiltration will be restricted and rainwater will be lost as runoff. The porosity of surface soil may have been reduced by clogging of pores with particles detached from soil aggregates under the impact of raindrops, or by the deposition of detached particles on the soil surface as impermeable crusts or seals.

The porosity of subsurface soil may be naturally low, or may have been reduced by compaction and tillage practices that have disrupted or destroyed pore spaces causing a zone of low permeability at the base of the tilled layer. The degree to which soil porosity is reduced by tillage is frequently sufficient to limit root penetration, but is less often so severe that permeability to water is significantly diminished. The overriding approach should be to instil in society, and in farmers, extensionists and researchers in particular, the will to create and sustain soil conditions that encourage the infiltration of rainfall where it falls, and to counteract the causes of runoff. This implies that the porosity of the soil must be at least maintained, or increased.

Capacity of the Soil Surface

Porosity of the soil surface is best maintained by first protecting it from the disruptive action of raindrops through a protective cover, usually of residues from the previous crop, a cover crop or mulch, and by ensuring the soil is not disturbed by tillage. The effects of conservation agriculture on higher infiltration and reduced runoff and flooding have been well documented in Brazil in particular.

If the whole concept cannot be applied immediately, improvements in soil moisture status of the soil can still be achieved, though probably not to the same extent, by other measures aimed at prolonging the useful life of rainwater. These include the use of surface residue covers alone, fallow periods under cover crops or natural vegetation, protection or temporary closure of grazing lands and forests from overgrazing, and operations on the contour, complemented by physical measures to detain rainwater.

The regular use of shallow tillage with disc or tined implements to break-up surface crusts to increase surface porosity and enhance rainfall infiltration is not recommended. The increase in surface porosity is only temporary and on crusting-susceptible soils tillage will need to be repeated after every rainstorm. Tillage leads to the disruption of pore spaces in the soil, and the use of discs, in particular, often causes compaction, which may impede root growth and rainwater percolation. Tillage also accelerates the loss of soil organic matter leading to a progressive deterioration of soil architecture and a reduction in the number and stability of pores that allow growth of roots and movement of rainwater. Regular tillage therefore is not recommended as a solution to restricted infiltration caused by low porosity of the soil surface.

A residue cover absorbs most of the energy of the raindrops that fall on it and by the time this rainwater reaches the soil below, its ability to disintegrate soil aggregates and detach fine particles is greatly reduced. Consequently, there is little or no clogging of surface soil pores by detached particles, and little deposition of soil particles that would form a crust on the surface. The benefits of a residue cover are most apparent on soils initially in reasonable physical condition, but even under these conditions runoff can sometimes occur despite a good soil cover. For example, runoff will occur when rainfall intensity is greater than the soil's infiltration rate, or when the soil's pore spaces are already filled with water because the soil is shallow, its water holding capacity is low, or its subsoil is only slowly permeable.

When a residue cover is applied to a soil with a very degraded surface of low porosity, the beneficial effect of the cover on infiltration may be initially limited. In such situations, it is advisable to accelerate the recuperation of surface porosity before applying residue covers by tilling the soil once to break-up the crust and any subsurface pans, followed by a fallow period under a cover crop to enhance the formation and stabilisation of soil porosity.

The choice of a cover material depends on what is locally available. Residue covers may consist of:

- Crop residues left in the field after harvesting the previous crop.
- Cover crops sown the previous season and left on the soil surface after slashing or applying a herbicide.
- Leaves and branches lopped from trees growing within the cropping area.
- Mulches of grasses, shrubs, weeds, litter, husks and other organic waste materials.

Improvements in crop yield and quality of soil occurred and a high proportion of farmers adopted these measures. Although erosion control was cited as the farmers' main reason for not burning crop residues any more, an important pointer to the benefits due to improved soil moisture conditions was evident in 1997. In that year there was a serious drought during the rainy season associated with the El Niño weather phenomenon. But according to the farmers, they were able to harvest almost as much maize as in a more normal year because of conservation of moisture in the soil as a consequence of the better soil status, while neighbours who had not adopted the system lost their crops to drought.

Nor did they lose their crops the following year during hurricane Mitch and the associated torrential rainfall, which caused disastrous flooding. The farmers noticed that the same mulch prevented the seeds from being washed away by rainstorms and facilitated rainwater infiltration so that they did not have problems of decaying plants during the heavy rains. A cross-check on this beneficial effect under the same extreme weather conditions comes from Honduras, where hurricane Mitch caused much erosive devastation on many hillsides, but less on those hillsides where soils were well protected by crop residues.

In a limited area of western Honduras the Quesungual traditional agroforestry system has been used by small farmers to produce maize, sorghum and beans. As rising population pressure makes the traditional slash-and-burn system increasingly unsustainable, there is increasing interest among farmers in the Quesungual system. It combines pruning of naturally regenerating indigenous trees and shrubs with normal agroforestry methods for growing high-value timber and fruit-trees. Before sowing, vegetation is cut down by hand without burning and is spread across the field together with the branches and leaves from pollarding. Crop seeds are then scattered or jab-planted with a stick through the mulch layer. Weeds are cleared infrequently, by hand or using a herbicide.

Mechanisms by which surface residue covers enhance rainwater infiltration. The physical contacts between a residue cover and the soil surface obstruct the movement of the runoff, slowing it down, giving more time for infiltration and so reducing the volume of runoff. Thus two aspects of surface cover can be distinguished:

— all surface cover absorbs the energy of raindrops and so prevents the loss of pore spaces into which rainwater can infiltrate;

— contact cover slows down any runoff, giving more time for infiltration.

The degree of contact cover is important especially on steep slopes, on soils with naturally low infiltration rates, and on degraded soils with surface crusts or seals of low porosity. Furthermore, it is the contact cover that is immediately accessible to soil macro-organisms and can stimulate their activity. Thus greater numbers of biopores are likely to be formed, leading to more rapid infiltration and percolation. This is why major disturbances such as tillage or incorporation of residues, mulches or other organic matter drastically reduces these positive effects.

Pliable materials of short length, such as leaf or grass mulch, which can be easily flattened by raindrops, will develop a high degree of contact cover and will substantially slow down the speed of runoff flow, generally resulting in reduced volumes of runoff. In contrast, inflexible long materials, such as woody branches of tall bushes that are not easily flattened by raindrops, will develop a low contact cover and so have less influence on the speed of runoff flow.

Advantages of Surface Residue Covers

The advantages of mulches are the same as for crop residue, i.e. increased infiltration, decreased runoff, and greater soil water availability. They both provide additional benefits, notably less soil water losses by evaporation, less weed incidence and water losses by transpiration, softer and more workable soils, increased earthworm activity, the incorporation of additional nutrients and frequently increased yields.

In western Kenya, mulching with Tithonia has given substantial yield increases of maize, kale, tomatoes and French beans. Net profits from mulching kale ranged from US$91 to US$1 665 per ha. In the semiarid zone west and north-west of Mount Kenya, maize yields increased by a factor of 4.4 when 3 t/ha of mulch were applied. Termites were not a problem in this area, probably because of the cool climate.

Disadvantages of Surface Residue Covers

The main disadvantage of applying mulch is the cost or labour of collecting, transporting and applying the mulch. This is not the case with crop residues, which are produced on-site. Often, there will be no suitable mulching materials in the vicinity of the farm, or there is insufficient labour available. Transporting large quantities of mulch for large-scale cropping is seldom economic and mulches cannot be applied after emergence to closely spaced crops. When a cover crop is used as mulch, there is the cost of slashing the cover crop or applying a herbicide. Similarly, lopping trees and distributing the branches and leaves over the cropping area requires considerable labour. On steep slopes, the application of residue cover's is not easy and requires much labour as well. Moreover, these materials are easily washed downhill on steep slopes.

Mulching materials and crop residues are often grazed by cattle belonging to the farmer, the community or the landowner (in the case of tenant farmers), fed to livestock, or sold as fodder. Sometimes these materials are in demand for thatching or fuel; in many semiarid areas they are rapidly consumed by termites, and in hot humid climates, they decompose rapidly.

Another disadvantage of mulches is a progressive decrease in soil fertility where the mulching materials are produced, unless manures or fertilizers are applied. In parts of Uganda, the residues of cereals grown on hillsides are used to mulch bananas on the lower slopes or valley bottoms, which become enriched in nutrients at the expense of the cereal areas. Soil erosion may also degrade the source areas when the cover provided by the vegetation is removed for use as mulch.

The use of soil covers is more common in subhumid and humid zones because of the greater availability of vegetative materials. Nevertheless, they are particularly suited to semiarid areas when materials are available and in the absence of severe termite problems. Mulches are often applied to limited areas of high-value horticultural crops and home gardens in easily accessible fields with gentle slopes.

Cover Crops Under Uncultivated Soil

When soils are so badly degraded that they must be taken out of production, soil porosity can be restored through the action of biological processes. This can be achieved by fallowing for 1 or several years under natural vegetation, natural vegetation enriched with fast-growing leguminous trees, or planted fallows. The accumulation of large amounts of biomass on the soil surface from the fallow vegetation associated with high biological activity and strongly developed root systems promote the biological recuperation of soil porosity.

Biological incorporation of residues into the surface soil results in higher soil organic matter in the upper few millimetres, which progressively extends into deeper layers overtime. The permanent cover of surface residues encourages soil faunal activity, which combined with higher soil organic matter contents leads to improved soil porosity.

A well-adapted, deep-rooting leguminous cover crop often speeds up the recuperation of soil porosity compared with a natural vegetation fallow because larger amounts of biomass are rapidly produced by the cover crop. Whereas a natural vegetation fallow may require 3-5 years, a cover crop may recuperate soil porosity in 1 year. When degraded soils are severely compacted, deep tillage with a subsoiler immediately prior to sowing the cover crop encourages establishment and development of the cover crop. If the degraded soil is severely deficient in phosphorus the application of P fertiliser will be necessary to encourage the establishment of the cover crop.

A constraint of soil recuperation by natural vegetation fallows in mechanised production systems is the problem of eliminating trees and excavating roots before returning to cropping. If a manual system is to be adopted, the problem is less serious. Herbaceous and shrubby cover crops can be eliminated much more easily by slashing, mowing or application of a systemic herbicide, and the subsequent crop may be sown directly into the residues of the cover crop.

Grazing Lands and Subsequent Protection

Low infiltration and high runoff can occur on grazing lands even on slopes less than 2 percent, as for example at Sebele, in Botswana. In this area, vegetation cover was considered to be the most important factor controlling infiltration and runoff, and catchments with a cover in excess of 70 percent generally had lower runoff compared with those with less than 70 percent cover. Although the percentage of grass cover in grazing lands has an important influence on rainfall infiltration, soil surface porosity can be more important, especially when overgrazing has degraded the soil, resulting in surface compaction and very low porosity.

On degraded grazing lands at Iiuni, Kenya, for example, even with 57 percent vegetative cover the runoff was in excess of 60 percent. The presence of algae growths on bare surfaces that were resistant to wetting encouraged runoff, whereas stone covers reduced runoff due to the creation of water-storage areas between the stones where the rainwater is detained, allowing more time for infiltration.

Forest provides an excellent protective cover made up of the canopy, low-storey bushes, herbs and surface litter, which combine to protect the soil surface from loss of porosity by direct impact of raindrops. The litter also serves as a food and energy source for soil organisms, which encourages the formation of soil organic matter and faunal passages leading to high infiltration rates. Where forests are not protected from grazing and litter is consumed by livestock, removed for use as mulch as in parts of Nepal, or is lost in fires, the surface cover may be diminished to such an extent that the soil becomes bare.

This is likely to be more serious under trees that discourage the growth of understorey herbs and shrubs, such as teak (Tectona grandis) and some species of Eucalyptus due to shade, high water use - especially by Eucalyptus - and to a lesser extent because of the acid nature of the litter. If the tree canopy is high, accumulated rainwater drops that fall off the leaves may be larger than normal raindrops and can fall with sufficient velocity to cause more damage to the soil than if there were no tree cover. This can lead to a pronounced loss of soil porosity, as can trampling by livestock, resulting in restricted infiltration and high runoff despite the high canopy cover.

Rainwater Drainage Zones

In regions where much of the rainfall occurs as light showers, the concentration of rainwater as near as possible to the crop will cause more of the rainwater to infiltrate deeply, where it is less susceptible to evaporation. In order not to lose this water by drainage beyond the crop's rooting zone and where there is no rooting restriction some solutions can be adapted, such as increasing the capacity of soils to retain water within the rooting zone, early planting to accelerate root development or changing to deeper-rooting crops.

The addition of large quantities of organic manure will increase the available water capacity (AWC) of soils and in theory this is a useful practice for reducing deep drainage losses. However, even in temperate climates the quantities of organic materials required to markedly increase AWC are very high, applications must be continued over many years and usually affect only the plough-layer depth. In tropical zones, where organic matter decomposition rates are much higher, the influence of organic manures on AWC is likely to be even less. Nevertheless, this practice may be feasible for small-scale farmers growing high-value crops where large quantities of organic manures and labour are readily available.

In low rainfall areas, it is frequently difficult to know when the rains have truly started, as initial rains are often followed by a dry period. Many farmers wait until the topsoil has been moistened to a depth of about 15-20 cm before planting, so that even if there is a subsequent short dry period there is sufficient water within the soil. However, this results in a delay in planting and for every day's delay yields will decrease, largely due to the loss of rainwater by drainage and evaporation, together with the loss of some released nutrients.

To overcome this problem and to allow crops to develop deeper rooting systems earlier on so that more of the rainfall can be utilised during the initial stages of the season, some farmers "dry plant" when soils are dry prior to the onset of the rains. To avoid premature germination before sufficient rain has fallen, the seeds are usually placed deeper than normal. Dry planting also has the advantage of spreading labour over a longer period.

Crops may also benefit from this practice by being able to utilise the nitrogen released at the start of the rains from the decomposition of soil organic matter, which reduces leaching and pollution of groundwater. However, there are a number of problems associated with dry planting, notably that some soils, and in particular hardsetting soils, are difficult if not impossible to till when dry. If seeds are not planted sufficiently deeply, they may germinate at the first rains and then die during a subsequent dry period.

Applying fertiliser to speed up crop canopy development and increase the shading of the soil surface will decrease the soil water lost by evaporation so that more is available to the crop. Planting crops equidistantly (i.e. with between-row spacing similar to within-row spacing) so that the soil surface becomes shaded more quickly would also be expected to reduce the proportion of soil water lost by evaporation. However, the effects of these agronomic practices on reducing evaporation losses will be much less than applying surface residues.

On permeable sandy soils that retain small quantities of available water for crop use, it is preferable to introduce deep-rooting crops that can utilise soil water at depth that would not be available to shallow-rooting crops. Examples of deep-rooting crops are almond, barley, cassava, citrus, cotton, grape, groundnut, olive, pearl millet, pigeon pea, safflower, sisal, sorghum, sunflower, sweet potato and wheat.

Climate on Agricultural Improvement

Traditional crop and livestock technologies have evolved such that climate and weather information becomes an integral component of the total system. The often complex timing required to fit operations into the seasonal cycle and to avoid specific weather hazards is codified in ritual and custom. An excellent example is provided by Stanhill, who shows a close correspondence between the timing and sequence of special prayers for rain in the Jewish liturgical calendar, established some 200 years ago, and the present-day dependence of wheat yields in Israel on the critical early part of the growing season. Careful analysis of such ritual and custom can yield valuable information about climatic patterns and weather hazards in regions where instrumental data are lacking.

The primary objective of most traditional systems is to minimise year-to-year variation in productivity and, especially, to minimise the risk of total loss of crop and/ or livestock. But this stability exacts a price in terms of potential yields foregone. While the traditional-crop cultivars and livestock breeds are hardy and resilient to the impact of climatic variation and weather hazard, they are usually low-yielding and incapable of an economic response to added inputs or new management techniques. So well adapted and finely turned are some of these systems that sudden change in any major component cultural, biological or physical can cause massive disruption

Modern, high-technology agricultural systems usually aim at maximising the expectation of profit or yield, using crop cultivars and livestock breeds that have been produced to fit specific environments and markets, and inputs of fertilisers, herbicides, pesticides and machine energy. Normally, buffering reserves of food and/or capital permit such systems to adopt higher-risk strategies that lead to much greater production on a time-averaged basis.

Acceptance of an occasional crop loss or uneconomic yield may be normal for a capital-intensive farmer, but it is unlikely to be acceptable to a subsistence farmer. These distinctions are important when globally assessing climatic impact on agricultural production. Estimation of risk is an important component of agricultural analysis, but the level of risk acceptance must be taken into account.

Evidence suggests that the earliest cultivators recognised that natural vegetation could provide an index of site quality for their crops. Modern man has added a little to this basic concept the transfer of information by analogy. The basic hypothesis is that all occurrences of a defined class of land will respond similarly to any imposed treatment. Climate, soil, vegetation and multi-attribute land classifications exemplify this approach.

These classifications then provide a basis for the selection of 'representative' experimental sites and for the subsequent extrapolation of results to other sites that share similar properties. Most current agrobiological research rests firmly upon such a foundation. Impact assessment can also employ analogues as simulators of future climate change. Analogues are sought that have the characteristics of the simulated or predicted climate, and the crop types or natural vegetation found within such climatic locations serve as the simulated or predicted vegetation.

Such a strategy is logical when the understanding of functional relationships between site attributes and crop response is qualitative or lacking, since this understanding is not required for classification. However, the better such knowledge is, the better will be the choice of attributes and the resultant classification. The thematic maps produced form a major component of an information system that is relatively inexpensive and highly portable. Usually agricultural research strategy is based upon climate and/or soil classifications, while forestry and pastoral research may place additional emphasis on vegetation classification.

Major deficiencies of the analogue approach reside in the static, multi-attribute character of the classifications upon which it is based and the implicit assumptions of covariance of attributes within mapped units. Compounding this is the location, season, cultivar and management-specific nature of the results from traditional agronomic experiments. For the individual farmer, the ideal location for an experiment is on his or her farm or as close as possible, since successful extrapolation of results is observed to hinge on proximity to the experimental site. Accordingly, there is continuing sociopolitical pressure to extend the network of research stations and experimental sites.

Given that agricultural research strategy will remain firmly based upon concepts of transfer of information by analogy for some time to come, there is ample scope for improvement and refinement of climate, land and soil classifications and for optimisation of experimental networks. Numerical taxonomic analysis and/or pattern analysis provide an array of classifications. The more relevant the attributes are, the better is the chance of a useful result. Commonly, the attributes measured represent a compromise between sampling density and site measurement.

Agroclimatic and bioclimatic classifications provide illustrations of the basic problems inherent in the analogue approach. While there is a considerable literature on bioclimatic classification and associated technologies, most of the general, multipurpose classifications are of limited use in solving specific problems. More direct coupling of climate data and specific crop response data offers prospects of more relevant classifications.

Agroclimatological zonation has always been particularly attractive to data-poor developing countries without an extensive network of meteorological and agricultural stations. They construct broad agroclimatological or agroecological zones using available temperature, precipitation, and soil and crop data. One example that illustrates this approach is the work of Jaetzold, who was asked by the Kenyan Ministry of Agriculture to provide a highly differentiated system of agroecological zones to help decide which variety of crops would best utilize the natural potential of a given district of arable land.

Multiple Regression Techniques

Since the pioneering work of Hooker and, more particularly, Fisher, both simple and multiple regression techniques have become standard practice in the agricultural and biological services, both for exploratory data analysis and for prediction. Such empirical/statistical modes of crop/weather analysis dominate the voluminous literature of agroclimatology.

Correlations have been established among a wide range of environmental variables and virtually all aspects of crop production. The utility of many such correlations under the conditions in which they were developed is not doubted, but the majority have limited application. Generally, the simpler the measure used, for example, total annual or seasonal rainfall, or annual mean air temperature, the more location-specific is the relationship.

Data on crop yield, temperature, and precipitation for a given district are the central components in statistical regression models. As reported by Waggoner, the model for a particular crop in a particular area is generally expressed as

$$Y_i = a + b_1 t_i + b_2 X_{2i} + \ldots + b_n X_{ni},$$

where

Y_i	=	estimated yield in the ith year;
a	=	intercept;
t_i	=	surrogate for technology in the ith year;
b_1	=	coefficient representing the effect of technology in quintals/hectare/year;
b_2 to b_n	=	coefficients representing the effect in quintals/hectare/ unit change in the weather;
X_{2i} to X_{ni}	=	weather variables such as precipitation, temperature, potential evapotranspiration (PET), and evapotranspiration (ET) in the ith year.

Thus, to quote Waggoner, 'by multiple linear regression the effect of weather, factor by factor, on the yields of crops is distilled from a history of weather and yields, accumulated for decades by faithful observers'.

Greatest success with this approach has been achieved where one, or perhaps two, environmental factors dominate crop performance. Thus, in high latitudes, temperatures that determine length of growing season may be most critical; in midlatitude, subhumid to semi-arid regions, rainfall amount may be most critical; and in the humid tropics, with high rainfall and near-continuous cloud cover, solar radiation receipts may be most critical.

The use of simple variables, such as monthly mean or seasonal mean precipitation and temperature, is common in large-area studies where the availability of data is restricted and/or computational limitations occur. Examples of successful applications of such analyses using simple data are provided by Thompson in his analysis of weather and technology in the production of wheat, corn and soybeans in the United States.

Usually, greater precision and generality are achieved if the primary climatic and/or soil data are transformed into indices that are more closely coupled to system performance. The derivation of crop/water stress indices from water-balance models has proved successful in the explanation of large-area variation in wheat yields. Subsequent developments in the understanding of crop response to water stress have led to the formulation of empirical/statistical functions for a very wide range of crops.

Inherent problems of the empirical/statistical approaches based on regression-type analyses relate to explicit assumptions of linearity of response, the interdependence of so-called independent variables, implicit assumptions that correlation implies causation, and, inevitably, the location, season, cultivar and management-specific nature of resultant relationships.

With care, these limitations can be minimised, and new techniques offer prospects of improved linear and nonlinear estimation procedures.

Regression techniques form part of the basic armory of methods used in the newer systems-based research strategies, both for exploratory data analysis and to quantify functional relationships between variables. Studies employing regression have been developed for a variety of questions about agriculture in relation to climate variability and change. As in the studies of Thompson cited above, a major objective has been the identification of systematic patterns of yield variation. For example, Newman and Waggoner have examined whether there has been an increase or decrease in crop yield variability between the 1930s and the present and whether there is a correlation in behavior between major grain-producing regions. In general, they find relative variability has decreased somewhat, while variability expressed as absolute yield has increased.

Regression-centered studies were extremely popular during the 1970s as nations and regions tried to assess their vulnerability to climatic variability. Good examples of national studies are found in McKay and Allsopp for Canada, National Research Council for the United States, and Takahashi and Yoshino for Japan and southeast Asia. For the World Climate Conference, Oguntoyinbo and Odingo explored relations among agriculture, climatic variability and various other causal factors in Africa, while Mattei, McQuigg, and Fukui prepared perspectives on the semi-arid regions, temperate regions and humid tropics, respectively.

Multivariate Techniques

These techniques represent a logical mathematical

development of regression-based analysis in that they permit simultaneous analysis of sets of interacting variables; multivariate analysis, factor analysis, principal component analysis, and principal coordinate analysis are examples.

While of principal value for data reduction and in exploratory data analysis, they can provide a basis for useful prediction and partial explanation. However, the derived principal component values, eigenvalues, or factor scores remain static representations of dynamic interactive processes, and their validity is strictly limited to the data domains used to develop them. While it is possible to determine the relative contributions of separate variables to the derived composite axes, it is often difficult to provide any physical explanation for their operation.

Sophisticated applications of multivariate approaches are found in the work of Monteith. Monteith developed a simple model of crop growth for tropical crops, including climatic factors, and has modified it for application in other climatic zones. For example, Monteith (employed the model to estimate the ultimate limit set by climate on the productivity of British farms, defining efficiency of crop production in thermodynamic terms as the ratio of energy output (carbohydrate) to energy input (solar radiation). Temperature and water supply were found to be the main climatic constraints on efficiency. Over most of Britain, where radiation and thermal climates are uniform, Monteith found rainfall to be the main discriminant of yield between regions.

Systems Analysis

Systems analysis is concerned with the resolution of any complex system into a number of simpler components and the identification of important linkages between them. Subsequent synthesis in the form of a symbolic representation (diagrams, flow charts) that is formalised as a set of logical statements or algorithms and mathematical formulae leads to the construction of a model of the real system. This model is then tested, using appropriate data inputs, and revised and developed to the point where it successfully simulates the behavior of the real system.

Beyond this point the model can be used to predict probable outcomes of experiments performed on the system. For many complex systems simulation-based experiments represent the only possible means of examining perturbations to the system. Climate and weather systems and biological systems are examples of complex systems that can benefit from the systems approach.

The process of formal analysis of any complex system yields other benefits in that it formalises what is known about the system, identifies the more important components and processes and significant constraints on performance, and provides a suitable vehicle for testing alternative management strategies and the 'what if ' questions posed by scenarios of global climatic change. Despite the proliferation of systems terminology and its ready incorporation into program titles and statements of research objectives, few agrobiological research strategies are truly systems-based.

Essentially, this is because a systems-research strategy demands interdisciplinary teamwork that transgresses the boundaries of long-established discipline and subject-specialist groups and competes with them for scarce resources. As in any research team, leadership is critical, but special problems relate to the maintenance of subject-specialist skills of team members, peer recognition, and the allocation of rewards and responsibilities. Most successful systems-research groups have evolved around a nucleus of a small group of committed scientists rather than through the imposition of a systems-research structure by higher authority.

Agroclimatic Synthesis

Past, present and, predictably, future modes of agroclimatic analysis and synthesis are inextricably linked with the broader goals of agricultural, pastoral and forestry development and land and water resource management. What, then, are these goals? Obviously, these will vary from field to field, farm to farm, region to region, nation to nation and continent to continent. But an ecologically sound and socially responsible goal would be to develop technologies that lift production, while maintaining the long-term stability of the land and water resource base. Implementation of such technologies rests, ultimately, with the individual land manager, whether profit maximiser or subsistence farmer.

The central problem, then, is how to prescribe a technology that is relevant to the land, labor, capital and management resources of the individual land manager. If it were possible to predict the performance of any production system at any location given a specified minimum set of soil/crop/weather-management data, then it would become possible to prescribe appropriate technologies. This ultimate objective is attainable, but that it requires a shift in emphasis away from reductionist/analytical research towards holistic/synthetic research. Of course, both types of research are necessary; they are complementary rather than competitive.

While much of current agrobiological research remains committed to the statistical differentiation of treatment effects on specific crops and cultivars, and on specific soils at specific locations in specific seasons, there are encouraging signs that greater efforts are being directed toward solution of the general problem. Of necessity, the earliest attempts at modeling agricultural systems were concerned with specific processes and/or specific products within restricted environments. Now, wide ranges of fully operational models of crop, livestock and forest systems are available or are under active development. Some, at least, are capable of general application across a spectrum of crops and environments.

Active development of simulation modeling and access to sophisticated computer technology are not in themselves, however, sufficient evidence of adoption of a systems-research strategy that aims at prescriptive rather than descriptive solutions. If, as stated, the primary objective is to develop models capable of providing both general solutions and location-specific prescriptions, then the focus must be on the balanced development of two interactive components (a) the models, and (b) the data base. The models should be fully accessible and operational at all times, yet capable of continuing improvement in logical structure and function.

The use of subsystem modules facilitates flexibility. The data base contains only those physical, biological, social and economic data that are specified as necessary for implementation of the model. Balanced development of both components is critical. Without access to prescribed input and test data, models cannot be implemented. On the other hand, the acquisition of datea for which no specific requirement exists can be wasteful of scarce resources.

2

Temperature and Humidity Control

Keeping products too cool can be a serious problem. It is important to avoid chilling injury, since symptoms include failure to ripen (bananas and tomatoes), development of pits or sunken areas (oranges, melons and cucumbers), brown discoloration (avocados, cherimoyas, eggplant), increased susceptibility to decay (cucumbers and beans), and development of off-flavours.

Cooling involves heat transfer from produce to a cooling medium such as a source of refrigeration. Heat transfer processes include conduction, convection, radiation and evaporation. If a ready supply of electricity is available, mechanical refrigeration systems provide the most reliable source of cold. Methods include room cooling, forced-air cooling and evaporative cooling. A variety of portable forced-air coolers have been designed for use by small-scale growers and handlers. However, a variety of simple methods exist for cooling produce where electricity is unavailable or too expensive.

Some examples of alternative systems include night air ventilation, radiant cooling, evaporative cooling, the use of ice and underground (root cellars, field clamps, caves) or high altitude storage. Ice can be manufactured using simple solar cooling systems, where flat plate solar collectors are used to generate power to make ice, which is then used to cool produce. Ice can be used either directly as package ice, to cool water for use in a hydro-cooler, or as an ice bank for a small forced air or room cooling system.

Several simple practices are useful for cooling and enhancing storage system efficiency wherever they are used, and especially in developing countries, where energy availability may be limited and any savings may be critical. Shade should be provided over harvested produce, packing areas, for buildings used for cooling and storage and for transport vehicles. Using shade wherever possible will help to reduce the temperatures of incoming produce and will reduce subsequent cooling costs.

Trees are a fine source of shade and can reduce ambient temperatures around packinghouses and storage areas. Light colours on buildings will reflect light (and heat) and reduce heat load. Sometimes spending money will save money, as when purchasing lighting equipment. High pressure sodium lights produce less heat and use less energy than incandescent bulbs.

Another aspect to consider when handling fruits and vegetables is the relative humidity of the storage environment. Loss of water from produce is often associated with a loss of quality, as visual changes such as wilting or shrivelling and textural changes can take place. If using mechanical refrigeration for cooling, the larger the area of the refrigerator coils, the higher the relative humidity in the cold room will remain. It pays however, to remember that water loss may not always be undesirable, for example if produce is destined for dehydration or canning.

For fresh market produce, any method of increasing the relative humidity of the storage environment (or decreasing the vapour pressure deficit (VPD) between the commodity and its environment) will slow the rate of water loss. The best method of increasing relative humidity is to reduce temperature. Another method is to add moisture to the air around the commodity as mists, sprays, or, at last resort, by wetting the store room floor.

Another way is to use vapour barriers such as waxes, polyethylene liners in boxes, coated boxes or a variety of inexpensive and recyclable packaging materials. Any added packaging materials will increase the difficulty of efficient cooling, so vented liners (about 5% of the total area of the liner) are recommended. The liner vents must line up with the package vents to facilitate cooling of the produce inside. Vented liners will decrease VPD without seriously interfering with oxygen, carbon dioxide and ethylene movement.

COOLING METHODS

Room Cooling

Room cooling is a relatively low cost but very slow method of cooling when electricity for mechanical refrigeration is available. When using room cooling, produce is simply loaded into a cold room, and cold air is allowed to circulate among the cartons, sacks, bins or bulk load.

This cooling method is best suited to less perishable commodities such as potatoes, onions, apples, sweetpotatoes and citrus fruits, since more highly perishable crops will deteriorate too much before being adequately cooled. Room cooling may be all you need if you handle chilling sensitive crops that need to be cooled from early morning harvest temperatures to storage temperatures of 10 to 13 °C (50-55 °F). The design and operation of cold rooms are fairly simple and no special equipment is required.

It is important to leave adequate space between stacks of boxes inside the refrigerated room in order for produce to cool more quickly. About 1 inch (2.5 cm) is sufficient to allow cold air to circulate around individual boxes. Produce in vented boxes will cool much faster than produce packed in un-vented containers. In many small-scale cold rooms, produce has been loaded into the room so tightly that cooling cannot take place at all, and despite the high cost of running the refrigeration system, the produce temperature never decreases to recommended levels.

Stacks of produce inside the cold room should be narrow, about one pallet width in depth (two or three cartons). Fans should be installed to move the cold air throughout the room. Air circulating through the room passes over surfaces and through any open space, so cooling from the outside to the center of the stacks is mostly by conduction. You'll want to monitor the temperature of the produce within the packages at various locations in the room to determine that the produce is being cooled as desired. Rearrange the stacks and measure the rate of cooling until you find the right pattern for your cold room.

Forced-air Cooling

Forced-air cooling pulls or pushes air through the storage containers themselves, greatly speeding the cooling rate of any type of produce. Many types of forced-air coolers can be designed to move cold moist air over the commodities. The example provided below is a fixed unit, where a fan is housed inside the wall of a cold room.

Cold Wall Forced-air Cooler

The canvas sheets must be well sealed, and pallet opening blocked for the tunnel style cooler to function properly. A portable forced-air cooler can be constructed using a canvas or polyethylene sheet. The sheet is rolled over the top and down the back of the boxes to the floor, sealing off the unit and forcing air to be pulled through the vents (vent area should be at least 5% of the surface area of the carton) of the cartons stacked against the cooler. This unit is designed to be used inside a refrigerated storage room. The fan unit is shown detached to illustrate how the air should flow within the cooler. For best results and minimum cost of operation, the warm exhaust air from the fan should be directed toward the return air inside the cold room.

Portable Forced-air Cooler:

Two types of forced-air coolers used for cooling cut flowers. Each is equipped with a fan to pull air from the cold room through the boxed produce. The recommended pattern of vents for cartons used to hold produce that is to be forced-air cooled. Vents should make up 5% of the total surface area, and should be located 5 to 7.5 cm (2 to 3 inches) away from the corners. A few large vents (1.3 cm =0.5 inch wide or more) are better than many small vents.

Hydro-cooling

Hydro-cooling provides fast, uniform cooling for some commodities. The commodity as well its packaging materials must be tolerant of wetting, chlorine (used to sanitize the hydro-cooling water) and water beating damage. The simplest version of a hydro-cooler is a tank of cold water in which produce is immersed. A batch-type hydro-cooler can be constructed to hold entire pallet-loads of produce. Conveyors can be added to help control the time produce stays in contact with the cold water.

Evaporative Cooling

These packinghouses are made from natural materials that can be moistened with water. Wetting the walls and roof first thing in the morning creates conditions for evaporative cooling of a packinghouse that is made from straw. The packinghouse is made with walls of wire mesh that hold charcoal. By moistening the charcoal with water each morning, the structure will be evaporatively cooled during the day.

Evaporative coolers can be constructed to cool the air in an entire storage structure or just a few containers of produce. These coolers are best suited to lower humidity regions, since the degree of cooling is limited to 1 to 2 C (2 to 4 F) above the wet-bulb temperature. A cooling pad of wood fibber or straw is moistened and air is pulled through the pad using a small fan. In the example provided here, O. 5 gallon of water per minute is dripped onto an 8 square foot pad, providing enough moist air to cool up to 18 crates of produce in 1 to 2 hours. Water is collected in a tray at the base of the unit and re-circulated.

An evaporative cooler can be combined with a forced air cooler for small lots of produce. Air is cooled by passing through the wet pad before it passes through the packages and around the produce. The air can be cooled to within a few degrees of the wet bulb temperature of ambient air.

Evaporative Cooler

The evaporative cooler is equipped with a vortex wind machine. Chicken wire was used to construct two thin boxes on opposite sides of the cooler that hold wet chunks of charcoal or straw. Water is dripped onto the charcoal or straw, and the wind turns the turbine, sucking moist, cool air through the load of produce inside the cooler. When using this cooler, temperatures are reduced to 3 to 5 °C (6 to 10 °F) below ambient air temperature, while relative humidity is about 85 %.

Evaporative coolers can be constructed from simple materials, such as burlap and bamboo. The "drip cooler" shown here operates solely through the process of evaporation, without the use of a fan. Cooling will be enhanced if the unit is kept shaded and used in a well ventilated area.

Drip Cooler

Two simple evaporative coolers have been developed and used in the Philippines for cooling and storage of vegetables (such as tomatoes, sweet peppers and mustard greens). Type a, is standing in a galvanized iron (GI) pan of water, and has another pan of water on top. The sides and top are covered with jute sacks kept wet by dipping their top and bottom edges into the pans of water.

In type b, the inner side walls are constructed from plain GI sheet with fine holes (spaced at 5 x 5cm) while the outer walls are made of fine mesh (0.32 cm) wire. The 1.5 cm space between the inner and outer walls is filled with rice hulls, kept wet by contact with a cloth that is dipping into the pan of water placed on top of the cooler.

Produce stored in these coolers has a longer shelf life than produce kept at ambient conditions. Tomatoes and peppers lost less weight and ripened more slowly, and could be kept for as long as they typically can be stored under refrigeration (about 3 weeks). Decay can be a problem, be can be controlled by washing in chlorinated water prior to cooling. Mustard greens lost much less weight and showed little wilting for up to 5 days.

An evaporative cooler located in the peak of a storage structure can cool an entire room of stored produce such as sweetpotatoes or other chilling sensitive crops. The vents for outside air should be located at the base of the building so that cool air is circulated throughout the room before it can exit.

The low cost cooling chamber is constructed from bricks. The cavity between the walls is filled with sand and the bricks and sand are kept saturated with water. Fruits and vegetables are loaded inside, and the entire chamber is covered with a rush mat, which is also kept moist. Since a relatively large amount of materials are required to construct this cold storage chamber, it may be useful only when handling high value products.

During the hot summer months in India, this chamber is reported to maintain an inside temperature between 15 and 18 °C (59 and 65 °F) and a relative humidity of about 95%.

Zero-Energy Cool Chamber

Night air ventilation: Storage structures can be cooled using night air if the difference in day and night temperature is relatively large. The storage facility should be well insulated and vents should be located at ground level. Vents can be opened at night, and fans can be used to pull cool air through the storeroom. The structure will best maintain cool temperatures during the heat of the day if it is well insulated and vents are closed early in the morning.

Chilling injury: Fruit and vegetable crops often are susceptible to chilling injury when cooled below 13 to 16 °C (55 to 60 °F). Chilling injury reduces the quality of the product and shortens shelf life. Symptoms often appear only after the commodity is returned to warmer temperatures, as when marketed.

Use of ice : Ice can be used as a bunker source of refrigeration (used by passing air through a bank of ice and then through the commodity) or as top ice (laid directly in contact with the product). Ice can cool a commodity only if it melts, so good ventilation is necessary for effective cooling.

Cross section - a gasoline or diesel engine must be mounted outside: Back of Room - an electric fan motor is commonly mounted inside the cold room, fan capacity should at least equal the empty volume of-the room (i.e. 12x8x8=768 cu ft.; so fan capacity should be 768 cfm at minimum.) (cfm= cubic feet/min)

Top view - vanes over the sub-ceiling greatly improve air distribution and hence cooling : Crushed or flaked ice for package icing can be applied directly or as a slurry in water. The use of ice to cool produce provides a high relative humidity environment around the product. Package ice can be used only with water tolerant, non-chilling sensitive products and with water tolerant packages (waxed fibreboard, plastic or wood).

Top ice is used for certain products during transport to help maintain a high relative humidity. Top ice can be used only with water tolerant, non-chilling sensitive products (such as: carrots, sweet corn, cantaloupes, escarole, lettuce, spinach, radishes, broccoli, green onions), and with water tolerant packages (waxed fibreboard, or wood). Top-ice on loads should be applied in rows rather than a solid mass. It is important not to block air circulation inside the transport vehicle.

Alternative Methods

Radiant Cooling

Radiant cooling can be used to lower the air temperature

in a storage structure if a solar collector is connected to the ventilation system of the building. By using the solar collector at night, heat will be lost to the environment. Temperatures inside the structure of 4 °C (about 8 °F) less than night temperature can be achieved.

Well Water

Well water is often much cooler than air temperature in most regions of the world. The water temperature of a deep well tends to be in the same range as the average air temperature of the same locality. Well water can be used for hydro-cooling and as a spray or mist to maintain high relative humidity in the storage environment.

High Altitude Storage

Typically air temperatures decrease by 10 °C (18 °F) for every one kilometer increase in altitude. If handlers have an option to pack and/or store commodities at higher altitude, costs could be reduced. Cooling and storage facilities operated at high altitude would require less energy than those at sea level for the same results. As a rule night ventilation effectively maintains product temperature when the outside air temperature is below the desired product temperature for 5 to 7 hours per night.

Increasing Relative Humidity

Refrigerated air tends to be lower in relative humidity than is beneficial for storage of most horticultural crops. The simplest method of increasing relative humidity of the storage air is to wet the floor of the room or mist the storage containers with cold water and allow the water to evaporate.

For a more permanent system of high relative humidity in the storage environment, moisture can be added to the refrigerated air. A fan draws air past the refrigerator's evaporator coils (R) then past wet moss or straw (M). The moist air is then pulled into the storeroom through a perforated wall (P).

Wet moss as a moisture source inside a refrigerated storeroom: Using a polyethylene liner in a fibreboard carton can help protect produce and reduce water loss in commodities such as cherries, nectarines, kiwifruits, bananas and herbs. Water vapour given off by the product is contained within the inner, increasing the RH around the product.

3

Pest Management

Integrated Pest Management (IPM) brings together into a workable combination the best strategies of all control methods that apply to a given problem created by the activities of pests. IPM has been defined in various ways but a more scientific definition describes it as, "the practical manipulation of pest populations using sound ecological principles to keep pest populations below a level causing economic injury". The emphasis here is "practical" and "ecological". There are many ways of controlling insect pests but only a few are practical, and fewer are ecologically sound, such that an undesirable citation is created.

Another term we frequently encounter is "intergrated pest control". It is offer used interchangeably with IPM, though in the strictest sense these terms are not identical. Originally, integrated control simply meant modifying chemical control in such a way as to protect the beneficial insects and mites, or integrating chemical and biological control methods. Subsequently the concept was broadened to include all suitable methods that could be used in complementary ways to reduce pest populations and keep them at levels which did not cause economic damage.

This essentially is IPM It includes a variety of options, any one of which may not significantly reduce the Pest population, but the sum total of which will give adequate reduction to prevent economic losses. A modern definition of IPM may be-the use of all available tactics in the design of a programme to manage, but not eradicate pest population so that economic damage and harmful environmental side effects are minimised.

IPM is not a static, unyielding system. It is dynamic, ever-changing, as we develop a better understanding of all factors that affect the system. These factors include climate, alternate host plants, beneficial insects and man's activities. In a narrow sense, IPM means the management of the few important pests generally found on our crops, but consciously or not it must include all insect pests, not only the "key" ones but also the secondary pests, which seldom do any harm. If this were not so, we might suddenly find some of these minor insect pests or even non-pests elevated to the status of serious insect pests because of our failure to consider them in the total scheme.

IPM as a concept is not new, but one that is receiving new emphasis as man looks for better methods to grow and store food for an expanding population, and at the same time preserve his environment. The rationale for using IPM is threefold. First, it can cut production costs mainly by reducing energy inputs. Secondly, IPM can reduce environmental contamination through the judicious use or reduced use of pesticides. And finally, an IPM programme allows for maximum utilisation of cultural practices and natural enemies and physical methods.

IPM can be designed to take advantage of the ecological principles governing pest population abundance. This requires a thorough understanding of the role of all the factors responsible for a pest population reaching certain levels at a particular time of the year, or duration of storage.

Elements of IPM

There are four basic elements of IPM: natural control, sampling economic levels, and insect biology and ecology.

The first element of IPM relates to the fullest utilisation of naturally occurring suppressive factors, including any practice by man which will make the total ecosystem less favourable for growth of the insect pest population. Obviously, this requires a thorough understanding of the ecosystem.

The naturally occurring suppressive factors may act directly or indirectly on pest populations. Indirectly, the ecosystem may be managed or altered in such ways as to make the environment more harmful to the pest and thus limit population growth. More directly, protection and the use of beneficial insects may help keep potentially damaging insect pest populations at subeconomic levels In storage, parameters that can be manipulated to control the buildup of a pest population are temperature, relative humidity, moisture content and composition of gases within the storage atmosphere.

The second element is that of using sound economic threshold (ETL) levels as the basis for applying control measures, especially chemical measures. Establishing and using dynamic ETL's provide a basis for delaying the use of insecticides. This permits the maximum utilisation of other control methods, such as the use of beneficial insects.

The use of economic threshold levels implies adequate sampling of all harmful and beneficial insects in the agroecosystem and particularly in any one crop at a given time. The levels found through sampling must then be measured against the economic level established for the crop, the beneficial insects, and the probable population trend of the pest species. The sampler thus becomes a key person in an IPM system.

The fourth element, insect biology and ecology, is essential to the fullest utilisation of the other three elements. Little concerning natural control can be understood without detailed knowledge of the biology and ecology of all the species present. This knowledge is also essential in establishing the role of each species in the system and in determining the amount of damage inflicted by each pest species. Adequate sampling is directly dependent a thorough familiarity of the species involved.

Knowledge of the biology of a certain problem pest will serve as a basis for planning the control strategies and provide operational guidelines for these strategies. It is important to know the relationship between the pest and the crop and the mortality factors, both biotic and abiotic which play a major role in the determination of pest population dynamics. An understanding of the sequential dominance of pests in relation to growth stages could provide the immediate impetus for developing a simple integrated control programme based on minimum pesticide application.

By delineating the succession of major pests at different stages of plant growth, the frequency, timing and dosage of insecticide application could be synchronised, hence avoiding pesticide use on a time-wise basis, or the "calender" method. The control programme could then be based on expected pest population at any given growth stage of storage duration.

Economic Levels

There is a wide variation in the degree to which pests

may be tolerated even for the same species in different areas, in different times of the year on different host plants, and in different stages of crop development. Thus, determination of the Economic Injury Level is critical in defining the ultimate aim of any pest management programme, and in delineating the pest population level below which damage is tolerable and above which specific intervention is needed to prevent a pest explosion and to avert significant damage.

Stern et al., defined Economic Injury Level as "the lowest pest population that will cause economic damage" while economic threshold level (ETL) of more accurately Control Action Threshlod (CAT) as "the density at which control measures should be applied to prevent an increasing pest population level from reaching the economic injury level". Although the damage or losses at the economic threshold can be tolerated of neglected, it is at this level that every effort should be made to reduce the pest population by various methods.

Determining the Economic Injury Level and Control Action Threshold is generally a complex matter based on detailed operations of pest ecology as it relates to bioclimatology, predation diseases, the effect of host plant resistance and the environmental consequences of applied control interventions. Rabb has suggested the following factors as essential for the determination of the Economic injury Level:

1. Amount of physical damage related to various pest densities.
2. Monetary value and production costs of the crop at various levels of physical damage.
3. Monetary loss associated with various levels of physical damage.

4. Amount of physical damage that can be prevented by the control measure.
5. Monetary value of the portion of the crop that can be saved by the control measure.

From this information, it is possible to determine the level of pest density at which control measures can be applied to save crop equal to, or exceeding the cost of control. The rise and fall of the Control Action Threshold is determined by the importance of the ecosystem, value of the crop, the pest status, and consumer standards. For example, Heliothis zea feeding on cotton boll, has an economic threshold of four larvae per plant and generally requires insecticidal treatment several times yearly. *H. zea* feeding on sweet corn, has an economic threshold approaching zero population in the United State since the consumer will reject sweet corn with any damage or one larva on it.

In special cases, where pests serve as vectors of plant, animal and human diseases, the economic threshold is zero. A single pest attack may cause the death of a valuable tree, a domestic animal or human. A good example is Aedes aegypti which transmits yellow fever. In grain storage, ETL is influenced by consumer attitudes, Export regulations often specify nil-tolerance of any live insects, be it injurious or beneficial.

Factors of IPM

Five general types of single component control methods may be used in IPM programmes in stored ecosystems. These are: chemical control, physical and mechanical methods, biological control, host plant resistance and regulatory control.

A variety of insecticides and acaricides have been and are continuously being developed for control of insect pests. However, these chemicals are but one tool and should be used in combination with other tactics in an IPM programme. The total reliance on chemicals has led to a crisis situation. However, IPM does not advocate the complete withdrawal of pesticides. That would be impractical. IPM simply demands use of pesticide only when necessary and at rates compatible with other strategies.

Physical and mechanical methods are direct or indirect measures that completely eliminate pests, or make the environment unsuitable for their entry, dispersal, survival and reproduction. Physical-mechanical control measures may include environmental manipulation, mechanical barriers, light taps, irradiation, thermal disinfestation, sanitation, etc. Many times, mechanical and physical methods require considerable extra equipment, materials and labour, hence, they may only be economical in certain situations. For field pests, these methods are rather inefficient but in a storage ecosystem, many of the physical techniques are effective and have great potential for use in an IPM system.

Biological control may be defined in a narrow sense as "the manipulation of predators or patho. yens to manage the density of an insect population". This definition does not include the naturally occurring control agents, but only parasitoids, predators and pathogens that are purposely manipulated by man. In a broader sense, it includes "the manipulation of other biological facets of the pest life system, such as its reproductive processes (i.e. sterile male technique), its behaviour, the quality of its food and so forth."

There are some constraints to the potential use and success of natural enemies. Predators, parasites and pathogens found amongst the grain will be regarded as contaminants by consumers and grain exporters. Thus, it makes it very difficult to maintain a pest population level that will enable the biological control agents to establish themselves. The use of pheromones is one of the potentially useful biological agents that could be utilised in IPM for monitoring and partially suppressing pest population not only in agricultural fields but in storage ecosystems.

The manipulation of the genetic make up of the host so that it is resistant to pest attack is called host plant resistance. Over the years there have been numerous successes in breeding for resistance to a variety of pests and currently many crops are being selected for this purpose. This approach has not been attempted to any great extent in stored products protection systems Investigations in this field have been few. However, research has provided evidence of the utility of varietal resistance in grain storage. Unless research on varietal resistance to storage pests is integrated with breeding of plants that are resistant to field insect pests and deceases the potential of this tactic in storage IPM is limited.

Fundamental regulatory control principles involve preventing the entry and establishment of foreign plants and animal pests in a country or-area, and eradicating, containing or suppressing pests already established in limited areas. Under the auspices of various quarantine acts, numerous control measures are implemented in an attempt to exclude potential pests, to prevent spread and to supplement eradication programmes. Ports of entry are the first line of defense against the introduction of new pests. Pests which break through the port of entry are eradicated or contained within limited areas. Quarantine action is used only against insects of economic importance, although it is sometimes necessary to contain insects which are of no economic importance in another country until their behaviour in a new environment can be studied

Trogoderma granarium is a most serious pest of stored commodities and every effort is extended to prevent its spread in international trade. In many countries, imported consignments found to contain *J. granarium* are segregated and immediately fumigated with methyl bromide. Lately, Prostephanus truncates originating from Central America has become a pest of international quarantine importance.

Each of the many methods in insect control has its place in IPM. There are many situations where two or more can be used in an integrated programme. Not all methods, however, are suitable for use in every situation. In a storage ecosystem, hygiene and good warehouse management are essential. It provides the framework for other supplementary infestation control methods. An IPM system would therefore supplement sanitation and good warehouse keeping with one or more combination of the following practices:

1. improved harvesting and threshing techniques
2. judicious use of residual insecticides
3. use of fumigants (MeBr; PH_3)
4. use of ambient aeration, and refrigerated aeration
5. atmospheric gas modification (hermetic; CO_2; N_2)
6. thermal disinfestation
7. irradiation techniques
8. insect resistant packaging
9. insect growth regulators: (IGRs: methoprene, hydroprene)

10. biological control (parasites, predators and entomopathogens, pheromones)
11. Use of resistant varieties if possible
12. Storage management (FIFO)
13. Adequate grain cleaning prior to storage storage.
14. Storage design (for pest exclusion, principally for rodont and bird pests)
15. Adequate grain cleaning prior to storage
16. Monitoring, evaluation and inspection of stored commodities, storage structures and their immediate surroundings.

Pest Management for Stored-Products

There is a number of differences for IPM for stored-products compared to field agriculture. There is far greater variety of tactics that could be employed for storage pest management. In both situations, a systems approach is used in order to facilitate monitoring and implementation.

Field Agriculture and Stored-Products

1. Ecological condition - more complex and dynamic.
2. Sanitation process - difficult to implement.
3. Economic threshold - appropriately applicable.
4. Physical control - often impractical and expensive.
5. Mechanical control - often impractical and expensive.
6. Regulatory control - more complex and require intensive logistical support.
7. Chemical control - considerable impact on the ecosystem.

8. Biological control - field
9. Host - plant resistance - whole plant modification relatively easier.

Stored Product Setting

1. Relatively simple and manageable.
2. Easier and alone could provide complete control.
3. Not generally applicable, food industry often require zero infestation.
4. Most physical factors could be manipulated under storage condition.
5. More feasible by as built - in feature of storage.
6. Detection and implementation generally self-contained.
7. More contained and limited in spatial proliferation.
8. Problem of contamination.
9. Genetic manipulation to render resistance to seed only more complicated.

In both situations three additional elements must be incorporated to the development of IPM: 1) appropriate people, 2) a systems approach, and 3) adequate evaluation. Aside from the individual expertise of the team members it is important that the people involved must develop a good personal and working relationship in the pursuance of the common objectives.

Economic Principles

Entomologists and biologists define pests as those organisms reaching a certain biomass or population level enough to cause physiological damage to crop and eventually impair either the quantity and quality of the produce. The concern used to be concentrated solely on devising control strategies with the aim of increasing the pest-kill efficiency and/or minimising the control's unwanted side effects on the environment.

In doing so,a damage function and a kill-function are determined simultaneously. Perhaps, the ultimate goal of these efforts is to realise the yield potential of crops by eliminating pest hazards which is possible only where there is maximum protection. In years past, maximum protection was closely associated with prophylactic dosages of pesticides.

Economists view pest problems and solutions quite differently. For one, pests are those organisms that destroy what man has produced, compete for resources needed by man in the production of food and fiber, or contaminate the environment in such a way that man finds it unhealthy or unattractive. An economist would not be very particular about the exact levels of pests, damages, and moralities; but would pay more attention to the value of pest damage and the costs of alternative control strategies with or without special consideration to time.

Usually, he would not desire for the yield potential of crops because he knows that it does not always pay to spend for maximum protection. In fact, the urge to control pests does occur to him until he finds that the cost of doing so is more than offset by the benefits. Biologists, to a great extent, do not worry about costs in as much as they find delight in quantifying apparent yield increases due to control action.

Almost for two decades now, attempts have been made to reconcile the disciplinary concerns of both the natural and social scientists to benefit the farmers and the public at large. To put their interests together, the US National

Academy of Sciences states that: entomologists develop and perfect various methods of insect control and determine their physiological impacts on crops; economists depend on entomologists for information regarding various inputs combined to produce a particular outcome.

To some extent, pest management was conceived to pool together the concerns of people coming from various fields in a most coordinated fashion. Pest management is defined as an ever-changing process of attacking pest problems by applying whatever economically feasible control of combination of controls that produces the best combination of immediate and long-term results in terms of both reduced pest damage and the absence of unwanted side effects. This implies that pest management may utilise a single-component-tactic or it may be accomplished through a complicated integrated pest management system as discussed earlier. Some Terms and Assumptions in Economic Analysis:

In the analysis of crop protection, it is necessary to define what are we putting in and what are we getting from this activity in order to evaluate if such is really worth it.

Input/Cost: The factors of production such as buildings, machines, materials, labour and management skills are called inputs and if valued at their respective prices are called farm costs or expenses. Let x = quantity of input and Px = Price of input. Then Cost = (Px)(X)

Output/Beturn: Through some physical or biological processes, the above inputs are transformed into usable products or output. Similarly, it can be valued by its price to get the return, revenue or benefit. The benefits from pest control is indicated by the increase in yield or added returns.

Let X = yield (increase) and Py = price of corp. Then
Return or Revenue = (Py)(Y)

Pest management is becoming more essential in our modernising agriculture as problems about pests and damages increase with time. As an input it has 5 components: pest control materials, labour and management or knowledge about pests, expected damage, and alternative control methods.

Pesticides: Pesticides used to be the single most important protectors of crops from insects, weeds and pathogens but because of the ever-increasing price and unwanted side effects, nonchemical methods are being tapped. Nevertheless, continued dependence on pesticides in the near future is a reality that pest managers have to reconcile with. Generally, pesticides form a regular feature of the farm cash expenses.

Non-chemical control: This method is likewise called labour control because it relies heavily on a farmer's manual efforts. Non-chemical methods do not necessarily mean that the materials needed for its utilisation are cost-free.

Efficiency: We define the most efficient pest control strategy as one that requires the least combination of labour cash input for a given level of yield increase, their factors held constant. The definition combines the "efficiency" and effectiveness concepts presented by Semple. According to him the efficiency of an insecticide is the minimum dosage required to reduce the population of target pests by a fixed proportion (80%) regardless of pest pressure.

Effectiveness, on the other hand, is indicated by the number of surviving insects of the extent of yield damage after the treatment. It is determined not by the "efficiency" of insecticides but by the strength of pest perssure as insecticides by the over-all farm operations. One needs to know the compatibility of his control measure to his pest problems and an effective manager does.

Optimum thresholds: At least, there are two sets of pest population thresholds found in pest management literature, one being a shadow of the other. Studies regarding plant physiology in relation to phytophagous organisms that reveal there is a tolerable level of plant destruction due to pests wherein the crop is still able to recover and remain vigorous.

Apparently, economic injury level (EIL), sometimes termed as the biological threshold level (BTL) is the critical pest density level going beyond what is tolerable to crops. With the EIL, biologists justify control actions. For the economists, such could never be unless the cost of control is zero. Economic threshold level (ETL) is the level of pest density at which suppression action is initiated, the control cost at least equals the value of expected damage.

Decay and Insect Control

High humidity in the storage environment is important for maintenance of high quality produce, any free water on the surface of commodities can enhance germination and penetration by pathogens. When cold commodities are removed from storage and left at higher ambient temperatures, moisture from the surrounding warm air condenses on the colder product's surfaces. A temporary increase in ventilation rate (using a fan) or increasing exposure of the commodity to drier air can help to evaporate the condensed moisture and to reduce the chances of infection.

Control of storage insects in nuts and dried fruits and vegetables can be achieved by freezing, cold storage (less than 5°C or 41°F), heat treatments, or the exclusion of oxygen (0.5% or lower) using nitrogen. Packaging in insect-proof containers is needed to prevent subsequent insect infestation. Some plant materials are useful as natural pesticides. Cassava leaves are known to protect harvested cassava roots from pests when used as packing material in boxes or bags during transport and short-term storage. It is thought that the leaves release cyanogens, which are toxic to insects. The ashes of the leaves of Lantana spp. and Ochroma logopur have been found to be very effective when used as a dust against aphids attacking stored potatoes.

The pesticidal properties of the seeds of the neem tree (as an oil or aqueous extract) are becoming more widely known and used throughout the world. Native to India, neem acts as a powerful pesticide on food crops but appears to be completely nontoxic to humans, mammals and beneficial insects. Any "natural pesticide" must be shown to be safe for humans before its approval by regulatory authorities.

Chemical Controls

Washing produce with chlorinated water can prevent decay caused by bacteria, mold and yeasts on the surface of produce. Calcium hypochlorite (powder) and sodium hypochlorite (liquid) are inexpensive and widely available. The effectiveness of the treatment will be decreased if organic matter is allowed to build up in the wash water. The effectiveness of chlorine increases as pH is reduced from pH 11 to pH 8, but at lower pH chlorine becomes unstable.

Fruits and vegetables can be washed with hypochlorite solution (25 ppm available chlorine for two minutes) then rinsed to control bacterial decay.

Alternatively, these commodities can be dipped in hypochlorite solution (50 to 70 ppm available chlorine) then rinsed with tap water for control of bacteria, yeasts and molds.

Sulfur: Sulfur is used on bananas as a paste (0.1 % active ingredient) to control crown rot fungi.

Sulfur dioxide: SO_2 is used as a fumigant (with a 10 ppm residue tolerance) on grapes to control Botrytis, Rhizopus and Aspergillus fungi. Careful calculation of the amount of sulfur dioxide required to treat grapes can greatly reduce the need to vent or scrub the storage air after fumigation to remove excess SO_2.

Sodium or Potassium Bisulfite: Bisulfites are usually contained within a paper or plastic pad that can be placed inside a carton to release SO_2 for control of molds on grapes during transportation and storage. The use of bicarbonate salts for prevention of postharvest decay has been used on fresh peppers, melons, potatoes, carrots and citrus fruits. These salts are very inexpensive, safe to use, readily available and accepted as "certified organic" and "chemical free" for marketing purposes.

Bicarbonate salts: Include Bicarbonate of soda, or "baking soda" ($NaHCO_3$), Potassium bicarbonate ($KHCO_3$)

Application methods (with or without chlorination as desired):

Method	Rate	Followed by:
overhead spray or dip	2% solution (use 2g in 100ml water or 20g per liter)	
overhead spray or dip	3% solution (use 3g in 100 ml water or 30g per liter)	rinse in water

Bacterial soft rot (Erwinia) of cabbage can be controlled by using lime powder or a 15% solution of alum (15g aluminum potassium sulfate in 100ml of water). After treatment of the butt-end of the cabbage heads, the produce should be allowed to dry for 20 to 30 minutes before packing.

On occasions when fungicides must be applied to produce, a simple tray with holes punched in the bottom can be used hold the commodity while it is sprayed. In the illustration below, a hand-operated knapsack sprayer is used to spray fungicides on bananas to the stage of runoff. The bananas can then dry in the perforated tray before further handling.

A back-pack sprayer is a useful tool for applying fungicides and other postharvest chemical treatments. A wide variety of sizes and types of sprayers can be purchased at farm supply centers or via companies on the internet.

When fruit is packed for export, fungicides are often applied to meet the requirements of international quality standards and to reduce deterioration during transport. The "cascade applicator" illustrated below was developed to apply fungicide uniformly and effectively by using a liquid curtain to drench the fruit.

Fruit in a perforated plastic tray is introduced on a roller conveyor belt (not shown) into the applicator. Inside a simple fan shaped deflector creates a curtain of liquid fungicide. The fruit passes under the curtain where it is drenched, then out of the applicator to drain on a tilted return tray. The tank holds up to 50 liters of fungicide solution, and a pump is mounted at the level of the tank outlet. A filter is fitted on the top of the tank to remove foreign matter from the return flow of fungicide from the applicator box and the return tray.

Fungicide Applicator

Some simple chemical treatments and handling practices can be used to prevent wilting and decay when handling and marketing fresh cut flowers, and can help cut flowers last much longer in the vase at home. If you are selling cut flowers, you could give a printout listing some of these tips to your customers.

- Start with a clean bucket or vase (washed with hot soapy water and rinsed thoroughly).
- Gently remove lower leaves so there will be no leaves in the bucket or vase water.
- Re-cut the stems, removing one or two inches at an angle under water.
- Use de-mineralised water if the water in your area is "hard" (creates white deposits on faucets) or has been "softened" (full of salts).
- Use warm to hot water (40 C or 100 F).
- Use one of the following solutions. They contain acid to improve water flow, sugar to help buds open and a preservative to reduce growth of fungi and bacteria.
  - mix 1 part lemon-lime soda with 3 parts warm water. Do not use diet sodas. Add ¼ teaspoon of household bleach per quart of water.
  - Use 2 tablespoons of lemon juice or lemon juice concentrate, 1 tablespoon of sugar and ¼ teaspoon of bleach in a quart of warm water. Add another ¼ teaspoon of bleach to the vase every 4 days.
  - Use a commercial floral preservative. These are inexpensive and convenient but may not be as effective as the two recipes provided above.

→ If you are using florist foam to arrange flowers, let it soak in the solution until it sinks on its own. Pushing it down into the vase will cause air bubbles to remain in the foam and can cause early flower death.

Cold Storage Treatments

Certain fungi and bacteria in their germination phase are susceptible to cold, and infections can be reduced by treating produce with a few days of storage at the coldest temperature the commodity can withstand without incurring damage (0 ° C for apple, pear, grape, kiwifruit, persimmon and stone fruits).

Rhizopus stolonifer and Aspergillus nigger (black mold) can be killed when germinating by 2 or more days at 0 ° C (32 ° F), and pathogen growth can be nearly stopped by storage at temperatures below 5 ° C (41 ° F).

Cold treatments can also serve to control some insect pests, and are currently allowed for the control of fruit flies, the false coddling moth, melon fly, pecan weevil and lychee fruit borer. Treatment for control of fruit flies requires 10 days at 0 ° C (32 ° F) or below, or 14 days at 1.7 ° C (35 ° F) or below, so treatment is only suited to commodities capable of withstanding long-term low-temperature storage such as apples, pears, grapes, kiwifruit and persimmons. Since requirements are subject to change, always refer to the latest edition of the APHIS treatment manual.

For produce packed before cold storage treatment, package vents should be screened to prevent the reinfestation by insects during handling.

Cold Treatment Protocol for Mediterranean Fruit Fly (Ceratitis capitata)

Temperature	Exposure period (days)
0 ° C (32 ° F) or below	10
0.6 ° C (33 ° F) or below	11
1.1 ° C (34 ° F) or below	12
1.7 ° C (35 ° F) or below	14
2.2 ° C (36 ° F) or below	16

Modified Atmosphere Treatments

For commodities that tolerate high CO_2 levels, 15 to 20% CO_2 -enriched air can be used as a fungistat to control decay-causing pathogens, such as Botrytiscinerea on strawberry, blueberry, blackberry, fresh fig and table grapes during transport.

Insecticidal atmospheres (0.5% or lower O_2 and/or 40% or higher CO_2) have been shown to be an effective substitute for methyl bromide fumigation to disinfest dried fruits, nuts and vegetables The effectiveness of insecticidal atmospheres depends upon the temperature, relative humidity, duration of exposure and life stage of the insect.

Low O_2 and/or high CO_2 have been used to kill certain insects in commodities that can tolerate these conditions. The effectiveness of insecticidal atmospheres depends upon the temperature, relative humidity, duration of exposure and life stage of the insect. The first five of eight examples are from Mitcham et al:

— The first and third instars of the greenheaded leafroller (Planotortrix excessana) and the first and fifth instars of the brownheaded leafroller (Ctenopseustis obliquana) and the light brown apple moth (Epiphyas postvittana) are completely killed in 2 months when apples are stored at 0.5 °C in 3% O2 and 3% CO_2.

— The eggs of the apple rust mite (Aculus schlechtendali) and the European red mite (Panonychus ulmi) are killed in 5.3 months when apples are stored at 2.8 °C in an atmosphere of 1% O_2 and 1% CO_2.

— Coddling moth larvae (Cydia pomonella) are killed in 3 months when apples are stored at 0 °C, 1.5-2% O_2 and less than 1% CO_2.

— In kiwifruit, the adult two-spotted spider mite (Tetranychus urticae) is killed by 40°C, 0.4% O_2 and 20% CO_2 in only 7 hours.

— When persimmons are stored at 20 °C, 0.5% O_2 and 5% CO_2, the third instar of leafrollers (Planotortrix excessana) is killed in 4 days and the larvae and adult mealy bug (Pseudococcus longispinus) is killed in 7 days.

— Sweetpotato weevil (Cylas formicarius elegantulus) has been controlled at ambient temperature in stored tropical sweetpotatoes by treatment with low oxygen and high carbon dioxide atmospheres. At 25 °C (76 °F), storage in 2 to 4% oxygen and 40 to 60% carbon dioxide results in mortality of adult weevils in 2 to 7 days.

— Coddling moth (Cydia pomonella) in stone fruits can be controlled at 25 °C (76 °F) by using atmospheres of 0.5% oxygen and 10% carbon dioxide for 2 to 3 days (adult or egg) or 6 to 12 days (pupa). Normal colour and firmness changes during ripening are not affected by treatment.

— Treatments with 45% CO_2 at 0 ° C (32 ° F) are being developed for several surface pests, including omnivorous leafroller (Platynota sultana), western flower thrips (Frankliniella occidentalis), and Pacific spider mite (Tetranychus pacificus) on harvested table grapes. This treatment requires 13 days at 0 ° tO_2 ° C (32 ° to 36 ° F) and could be conducted in a marine container during transport. However, neither this nor other insecticidal CA treatments are approved yet as quarantine treatments.

Heat Treatments

Hot water dips or heated air can be used for direct control of postharvest insects. In mangoes, an effective treatment is 46.4 °C for 65 to 90 minutes, depending on fruit size, variety and country of origin. Fruit should not be handled immediately after heat treatment. Whenever heat is used with fresh produce, cool water showers or forced cold air should be provided to help return the fruits to their optimum temperature as soon as possible after completion of the treatment.

Some pathogens are susceptible to heat treatments. Brief hot water dips or forced-air heating can be effective for disease control, especially for reducing the microbial load for crops such as plums, peaches, papaya, cantaloupe and stone fruits, sweetpotatoes and tomatoes.

Biological Control Products

Two biological control products (antagonistic organisms) are currently used as complementary tools (to chemical and/or heat treatments) for the management of post harvest decays together with other strategies as part of an integrated pest management programme for a few fruits and vegetables.

Two plant growth regulators can be used to delay senescence of citrus fruits and consequently delay their susceptibility to decay.

4

Deciduous Fruit Production

The poor agriculture with its low level of diversification is no longer sufficiently interesting as an economic activity to attract the people who live in those areas. Consequently, many of them abandon the hill and mountainous areas, migrating to the towns with the hope for a better living. This leads to a deterioration of the status of agriculture, not to mention the damage to the natural environment resulting from neglect.

By developing deciduous fruit trees in the highlands the farmers' income can be substantially improved and erosion problems can be reduced. Deciduous fruits tend to be high value crops and in the tropics are highly marketable. However, on the one hand many technical problems related to the development of deciduous fruits such as apple, pear, peach, plum, apricot, persimmon etc., need to be solved. On the other hand the results already obtained in India, Indonesia, Pakistan, Nepal, Bhutan, China and of course Thailand merit dissemination.

Many countries in Asia have physiographic features and climatic conditions that could be effectively exploited for the production of deciduous fruits. However, the production systems that are already in place and which have been developed by farmers themselves are comparatively inefficient and less competitive. These traditional systems are no longer economically viable. Whilst many farmers in the highlands have held on to these traditional systems as a means of survival, others have abandoned these areas and moved in search of better prospects.

Vast areas that have been abandoned have been subject to environmental degradation through neglect. Many governments have resorted to interventions such as re-afforestation to stabilise these areas. Whilst these attempts have helped to restore the ecological balance, the problem of depopulation has not been arrested and most countries are grappling with the problem of resettling people in these areas and improving the conditions for farmers to develop agriculture on a settled basis.

Deciduous fruits such as apples, pears, peaches, plums apricots, nectarines, persimmons etc., seem to adapt well to the highland areas. These crops also help to stabilise hill slopes and arrest environmental degradation, even without resorting to expensive bench terracing, which has been the traditional system in the hill and mountainous areas.

Transport is often difficult from some of the inaccessible areas, but these fruits tend to have better shelf-life and storability than most tropical fruits and are also more amenable to drying. Other possibilities for the producer are simple fruit preservation techniques in order to help better utilisation and marketing. Once these fruits reach urban markets, there is ready acceptance as they are non-traditional commodities much sought after by consumers.

Highland areas of the tropics are characterised by steep slopes and highly indented valleys that have poor road accessibility and communications. Most areas support populations that have few avenues for development, which often results in abandoning the land and migrating to lower elevations and urban areas which offer alternate avenues for survival. There is also a lack of programmes for such regions as they are not well patronised by governments or by the private sector. A well-organised development plan for the production of deciduous fruits would help generate more income through the supply of these non-traditional commodities to markets within each country and can even substitute expensive imports.

It must be borne in mind, however, that there are many production technology gaps for the successful cultivation of deciduous fruits in tropical highlands. This is a challenge for the production specialists who are called upon to cater to the needs of every crop and variety.

The first step in the process of deciduous fruit development would be to identify superior cultivars of quality fruit which are market demanded, and suitable for different growing areas in each country. These varieties should be vegetatively propagated using appropriate rootstocks. Availability of high quality planting material is of paramount importance in the whole process.

Similarly, appropriate crop management practices such as training and pruning of the plants, fertiliser application, weeding, mulching, supplementary irrigation, pest and disease control etc. can play a very important role in increasing the production and productivity of these crops.

It should be noted that plant protection is an important component of the crop management programme. However, in order to develop an eco-friendly package, it will be necessary to reduce dependence on chemical control methods and to switch over to integrated pest management practices. Processing and marketing issues are equally important and need to be given due attention. There is also a need to strengthen advisory services to guide the agricultural communities in the proper adoption of new innovations.

Countries in the region have highland areas with similar climates. It must not be ignored that many agricultural communities have developed several useful technologies on their own. These useful technologies that have evolved over a period of many years should also be given due recognition. What is now needed is to assimilate this information, refine the technologies where and when necessary, and share such information among countries. This will help expansion of deciduous fruit culture to other areas where diversification and development are needed.

Fruit Production in India

India produces all deciduous fruits including pome fruits and stone fruits in considerable quantity. These are mainly grown in the North-Western Indian States of Jammu and Kashmir, Himachal Pradesh and in Uttar Pradesh (U.P.) hills. The North-Eastern Hills region, comprising of the States of Arunachal Pradesh, Nagaland, Meghalaya, Manipur and Sikkim also grows some of the deciduous fruits on a limited scale.

Due to introduction and adaptation of low chilling cultivars of crops like peach, plum and pear, they are also now being grown commercially in certain areas of the north Indian plains. Out of all the deciduous fruits, apple is the most important in terms of production and extent.

Apple was introduced into the country by the British in the Kullu Valley of the Himalayan State of H.P. as far back as 1865, while the colored 'Delicious' cultivars of apple were introduced to Shimla hills of the same State in 1917. The apple cultivar 'Ambri', is considered to be indigenous to Kashmir and had been grown long before Western introductions.

Pears and other deciduous fruits were domesticated successfully in the early part of the 20th century, although some of them were reported to occur under semi-wild conditions much earlier. Apricot was found growing in the drier pockets of north-western Himalayas and two apricot varieties, locally known as 'Halman' and 'Rakchaikarpo' are reported to be indigenous to Leh-Ladakh areas of J&K State.

Sweet cherry was introduced from Europe before India's independence in 1947, while commercial cultivars of sour cherry have been brought mainly from USA in more recent years. The European and Japanese plum varieties are grown both in high and low hill areas. A plum variety 'Santa Rosa' reported to be a hybrid between Japanese and American species predominates plantations in the hills. Low chilling cultivars of peach and nectarine such as Flordasum, Flordared, and Sunred nectarine are successful introductions to the north-Indian plains. Some local selections of peach, plum and sand pear are also cultivated on a commercial scale in sub-tropical - marginal chilling areas of north India.

Deciduous Fruit Crop Cultivation

India grows crops like apple, pear, peach/nectarine, plum (Prunus domestica L.), apricot (Prunus armeniaca L.), sweet cherry (Prunus avium L.) and sour cherry (Prunus cerasus L.) on a commercial scale.

The promising cultivars of different temperate fruits in 3 major deciduous fruit growing States of India. Some relevant information on cultivars and cultivar selection is indicated below:

Apple: Over 700 accessions of apple, introduced from USA, Russia, U.K., Canada, Germany, Israel, Netherlands, Australia, Switzerland, Italy and Denmark have been tried and tested during the last 50 years. The delicious group of cultivars predominates the apple market.

The areas covered under Delicious cultivars are: 83% of the area under apple in H.P., 45% in J&K and 30% in U.P. hills. In more recent times improved spur types and standard color mutants with 20-50% higher yield potential are favored. The important selections are:

— Spur types - Red spur, Starkrimson, Golden spur, Red Chief and Oregon spur.

— Color mutants - Vance Delicious, Top Red, Skyline Supreme.

— Low chilling cultivars - Michal, Schlomit.

— Early cultivars - Benoni, Irish Peach, Early Shanburry, Fanny

— Juice making cultivars - Lord Lambourne, Granny Smith, Allington Pippin.

— Scab resistant cultivars - Co-Op-12, Florina, Firdous, Shirean.

— New Hybrids - Lal Ambri, Sunehari, Amred, Chaubatia Anupam & Chaubatia Princess developed in India.

In H.P. monoculture of a few cultivars such as Royal Delicious, Red Delicious and Richared have started showing negative impact on the apple industry. Serious problems of apple scab disease and outbreak of premature leaf fall and infestation of red spider mite are causing great concern. U.P. Hills, particularly the Kumaon hills division, have the unique advantage of early harvest of apple, mainly due to cultivation of early maturing varieties like

Early Shanburry, Fanny and Benoni. The early maturing varieties are harvested 2-3 weeks before the arrival of fresh apple from H.P. and J&K, and hence fetch very remunerative prices.

Pear: In pear for higher altitude conditions high chilling requirement varieties (like Bartlett) are mainly grown. In more recent years, red-color strains of pear like Max Red Bartlett, Red Bartlett and Starking are replacing yellow colored cultivars. In warmer sub-mountainous areas of H.P. and sub-tropical Punjab oriental pear cultivars like Baghugosha, Kieffer, China and sand pear Patharnakh are cultivated commercially both for table and processing purpose.

Apricot: Generally there are two types of apricot, namely, sweet kernel type and bitter kernel type. About 81 exotic accessions and 19 indigenous cultivars were collected and evaluated. The local types Halman and Rakhaikarpo have been popular, whereas exotic introductions namely, Nari, Kaisha, Shakarpa and New Castle are promising. These cultivars are recommended for dry cold areas. The USA variety Nugget is self-fruitful and bears sweet and attractive colored fruits.

Peach: For colder conditions the peach cultivars July Elberta, Elberta, Peshwari, Quetta, Burbank and Stark Earliglo are well adopted. Low-chilling cultivars viz. Flordasum, Flordared, Shan-e-Punjab, Sharbati and Sunred (nectarine) have become popular in subtropical belts of U.P. and Punjab States.

Plum: A large number of cultivars (283) have been introduced into the country. European plums performed better in the hills, while Japanese plums are more adopted in sub-mountainous lower elevations. Leading cultivar in the hills is Santa Rosa. In the north-Indian plains small fruited cultivars like Titron, Kala Amritsari, Kelsey, and

Alubukhara showed better performance. A good number of low-chilling Florida hybrids Sungold, Redgold etc, are under evaluation.

Cherry: Many cultivars of sweet cherry have been introduced from Europe, USSR and British Columbia. Promising exotic cultivars like Bigarrean Napoleon, Black Heart, Guigne Noir for J&K and cultivars like Black Tartarian Bing, Napoleon (white) Sam, Sue (White), Shella for H.P. have been identified. For warmer climate, cultivars like Summit, Sunburst, Lapins, Compacat and Stella have been found to be promising.

Rootstocks

— Apple has been propagated mainly on seedling stocks of apple (Malus pumila Mill). Clonal rootstocks of M and MM series were introduced from East Malling Research Station of U.K. and M-9, M-26, M-4, M-7, MM-106 and MM-111 were identified as promising. Merton 779 was recommended as a commercial rootstock for apple for Kumaon hills of U.P. However, clonal rootstocks have not found favour among the apple growers. In-vitro protocols of M-9, MM-106 and MM-111 have been developed. Clonal rootstocks have been preferred only for high density orchards. Four rootstocks, namely, EMLA-111 for drought prone areas, EMLA-7 for slopy land and EMLA-106 for slopy and less clayey soil and EMLA-9 for high density planting on irrigated deep soils have been recommended. In pears, seedling stocks of Pyrus pashia (Kainth) is commercially used. Clonal rootstock Quince A with semi-vigorous growth has some promise. Newer clonal rootstocks like Old Home X Farmingdale series and BA 29C are under evaluation.

— For apricot, seedlings of wild apricot have been found to be good giving a smooth union and vigorous scion growth. Peach can also be tried as a rootstock, particularly for dry and light soils. Myrobolan plum is recommended for high moisture soil conditions.

— Peach is normally grafted on wild peach seedlings, plum (cv. Kabul Greengage) and even on apricot. Wild peach produced healthy and high-yielding plants. For controlling tree size, apricot rootstocks which impart dwarfing may be used.

— For plum, wild peach and wild apricot are commonly used as rootstocks. Among clonal rootstocks, Myrobolan B. was found to be most promising. Wild apricot seedlings and wild peach were also found to be suitable rootstocks for commercial cultivars like Santa Rosa, Mariposa and Beauty.

— Commercial plantations of sweet cherry were raised mainly on Mazzard and Mahaleb rootstocks. Seedlings are raised from late maturing cherry cultivars and stones of Waterloo or Misri cultivars are often used as rootstocks. Rooting of cherry suckers has been made possible with use of rooting hormones like IBA (6000 ppm) in combination with NAA (2000 ppm).

Area and Production

Deciduous fruits are mainly cultivated in North West Hills Region of India, comprising of States of Jammu & Kashmir (J&K), Himachal Pradesh (H.P.) and Uttar Pradesh (U.P.) and in the North Eastern Hills Region in the States of Arunachal Pradesh, Nagaland, Meghalaya and Manipur.

Apple

Although there has been 5-6 fold increase in apple production during the last 50 years, the productivity level is still very low (5.56 t/ha). Apple cultivation received greater attention by the growers. In H.P, area under apple increased from 3026 ha in 1960-61 to 78296 ha in 1995-96 with a corresponding increase in yield. J&K covers about 78007 ha under apple with a production of 714834 tons.

In the U.P. hills (8 districts) apple occupies about 30 per cent of the area under fruits and contributes 46.9 percent of fruit production. The area covered under apple in U.P. hills is 55200 ha with production of 210000 tons of fruits. In the North-Eastern Hills Region, good quality apple is produced only in the rain-shadow belts of Arunachal Pradesh (5523 ha), and in Nagaland a very small area (64 ha) has been brought under apple cultivation.

About 99 percent of India's apple area falls under the North Western Hills region, covering 6 districts of J&K (Srinagar, Budgam, Pulwama, Anantanag, Baramullah, Kupwara), 6 districts of H.P. (Shimla, Kullu, Sirmour, Mandi, Chamba, Kinnaur) and 8 districts of U.P. (Almora, Nainital, Pithauragarh, Tehri, Pauri, Chamoli, Uttarkashi, Dehradun). In the North-eastern Hills region, good quality apple is grown in a small area in Tawang belt of Kameng district in Arunachal Pradesh.

The Tawang area is basically a rainshadow belt and therefore, permits a longer period of sunshine and freedom from heavy rains, making it ideal for apple. Apple is also grown in Sikkim and Nagaland but the production is not a major success. Presently, a small quantity of apple produced in India is exported, mainly to Bangladesh and Sri Lanka.

Pear

Pear is grown under a wide variety of climatic regimes, ranging from cold dry temperate hilly conditions to warm humid subtropical conditions on the plains of northern India. Its cultivation is most extensive in J&K State (14012 ha), followed by U.P. (10550 ha). In the plains of Punjab, pear is grown in a specific area (7899 ha), whereas in H.P, it is widespread from the foothills to the highlands (600-2700 m) in areas experiencing 500-1500 chilling hours.

Others

There is no reliable data on area and production of other deciduous fruit crops. Stone fruits as a group occupy an area of 0.11 million ha, with 0.14 million tons production. In H.P about 28 thousand ha are covered under peach, plum, apricot, almond and cherry. In the North-western Himalayan region, peach holds greater promise because of its utilisation for canning purposes. Peach is grown mainly in low and mid hilly areas (1000-2000 m above msl), except the low chilling cultivars belonging to the Florida group, which can be grown very well under sub-tropical conditions.

European plums require more chilling than Japanese plum. Japanese plums are generally grown in low and mid-hilly areas (1000-6000 m) conditions. The climatic requirement of apricot is almost the same as that of plum, while cherries require a colder climate (1000 chilling hours below 7° C). Cherry is commercially cultivated in J&K where the hailstorm problem is not encountered.

In the North-Eastern Hills (NEH) Region, these crops can be successfully grown in Arunachal Pradesh (Kameng, Siang, Tirap and Lohit districts), Meghalaya (Central plateau of Khasi & Jaintia hills), Manipur (Maram, Tadubi, Mou, Ukhrul) and Nagaland (Mokokchung, Wokhla, Thensang, Kohima, Phek districts). The productivity of all the stone fruits is low and estimated yield of peach is 0.50, 0.73, 2.11 and 2.23 t/ha in H.P, J&K, U.P, and NEH Regions respectively. The productivity of cherry in J&K is approximately 1.73 t/ha, while average yield of apricot is 0.42, 0.20 and 0.28 t/ha in H.P, J&K and U.P. hills respectively.

Production of Planting Material

Nurseries

There is a large number of government and private nurseries engaged in multiplication of planting material of deciduous fruit crops. In addition, the State Agricultural Universities and the Research Institutions multiply planting material of improved cultivars for sale and distribution to the farming communities. The Government of India had supported establishment of a large number of fruit nurseries, both in public and private sectors as planned activities.

During the 8th Five Year Plan it is estimated that over 55 million nursery plants of different perennial crops, including temperate fruits, have been produced and distributed under the scheme. There was a target for the establishment of 85 big nurseries, 587 small nurseries, and 37 tissue culture units (20 by the private sector and 17 by the Government) of different fruit crops. Exact number of these nurseries engaged in production of pome and stone fruits is not known. However, the existing nurseries (more than 600), covering both public and private Institutions, are sufficient to meet the requirement of planting material of deciduous fruit crops.

Rootstock Information

In all the pome and stone fruits vegetative propagation techniques of budding or grafting are followed for multiplication of planting material on standard rootstocks, raised both from seeds or through clonal methods. Cropwise details are as follows:

Apple

For raising rootstock seedlings, seeds of crab apple or commercial cultivars are stratified during December for

2-3 months at 2-5°C. One year old seedlings are used for budding/grafting.

Clonal rootstocks are raised through mound or stool layering. The mother plants are allowed to grow for one season and cut back to 3 cm from the ground level just before the growth begins. When new growth is about 10 cm, the shoots are covered with soil leaving the growing parts exposed. Rooted layers are cut off close to the ground level and planted in nursery beds for grafting/budding.

In H.P. hills, different types of budding/grafting are recommended:

— Chip budding in mid-June and mid-September

— Whip and Tongue and cleft grafting in February - March

— T-budding in May-June

Pear

Seedlings of Kainth (Pyrus pashia Linn.) or Shiara (P. serotiana Rehd) are stratified in moist sand for 35-45 days at 2-5°C. Tongue grafting in February-March or T-budding during June-July is recommended.

Others

Plum and apricot are generally grafted on wild apricot seedling rootstocks. Peach is also used as a rootstock. Peach is budded/grafted on wild peach and peach-almond hybrid rootstocks. Hard wood cuttings (treated with 500 ppm IBA) can also be employed for raising rootstocks. For apricot, peach, plum and cherry shield budding during July-August has been recommended for the hills of J&K.

Micro-propagation through tissue culture technique has not yet been commercialised for temperate fruits, although tissue culture protocols have been developed at research Institutes/Universities. Colt, a rootstock for cherry can be multiplied in large numbers by following the tissue culture technique. Shoot buds are cultured in MS medium containing 1-2 mg/liter benylaminopurine. Individual shoots are rooted on MS medium containing 1 mg/liter Indole Butyric Acid (IBA). The rooted plants are transferred to sterilised sand + soil (1:1) mixture and covered for 7-10 days for hardening.

Establishment of Deciduous Fruit Orchards

Fertile sandy-loam to clay loam soils with pH range from 5.5 to 7.5 and free from water logging conditions are suitable for establishing deciduous fruit orchards.

In flat lands/valleys, square, rectangular or triangular planting may be adopted. In hill slopes, planting on contours or terraces is recommended. In shallow slopy lands, small terraces (half-moon terrace), may be made to establish the plants and large scale disturbance of surface soil need to be avoided.

Planting of deciduous fruit plants is done during winter months from the end of December to mid March. Planting is done in the center of pits (1x1x1m cube or circle), prepared a month before planting. While refilling the pit, 50g of aldrin powder (15%) is mixed with the soil.

After planting, the young plants are supported with stakes and basins are kept free of weeds. Mulching with dry grass or polyethylene is advisable. Irrigation is provided after planting.

Spacing varies from species to species and depends also on the type of rootstocks used. For apple raised on seedlings of crab or other commercial varieties of apple, a planting distance of 6x6m or 7x7m accommodating 277-205 plants per ha is recommended for J&K hills. On clonal rootstocks like M4, M7, M26, MM106 a spacing of 4.5m x 4.5m (555 plants/ha) is suggested.

For spur type varieties and standard colored mutants of apple, high density planting on dwarfing rootstocks like M9, M4, M7 and MM107 has been found to be feasible. Fruit yield of 30-35 tons/ha has been achieved in 12 year old orchards of color mutants of apple on MM106 under a planting density of 2222 plants/ha (3m x 1.5m) in cooler hills of H.P.

Management of Orchards

Modified central leader system of training has been recommended for both apple and pear on seedling rootstocks. The proportional heading back and thinning out system of pruning is adopted after the juvenile phase of plant growth. Spur pruning encourages vegetative growth and helps in new spur development in old plantations. For high density planting on semi-dwarfing and dwarfing rootstocks spindle bush, dwarf pyramids and cordon system of training are suggested. Spindle bush on M7 and modified central leader on MM106 rootstocks are successful.

Peach is generally trained to open center system, while in sweet cherry modified central leader system is followed. Pruning is done in such a way so that 20-50 cm new growth in young trees and 25-30 cm in older trees are secured every year. In sweet cherry about 10% of fruit bearing area should be removed annually, whereas in apricot slightly heavier pruning should be practiced to keep the spur system renewed.

Nutrient requirement varies from place to place. In Himachal Pradesh widespread deficiencies of N, P, K, Ca, Mn, Zn and B have been recorded. A fertilizer dose of 700:350:700 g of N, P and K for full grown bearing apple tree has been recommended. It has been observed that VAM (Vesicular arbuscular mycorrhizea) fungi increases P uptake by apple roots.

In apple orchards, grass mulching (10cm thick) with one application of post-emergence herbicide Glyphosate (0.8 kg/ha) was found to be effective. The shrubby weeds were, however, best controlled by 500 ppm Gramaxone + 100 ppm 2,4,5 - T application. Moisture conservation was maximum under grass mulching and, therefore, recommended for apple in H.P. Mulching with oak leaf was effective in U.P.

Trickle irrigation at 75 per cent of field capacity results in better tree growth and higher fruit yield in apple when raised on semi-dwarfing rootstocks. For apple, most critical period of water requirement is April to August and peak water requirement is just after fruit set. 114 cm of water during whole year through 19 irrigations have been recommended.

Drip irrigation saves water considerably. Field trials indicated a total irrigation requirement of 3840 liters water per tree under conventional system of irrigation; under drip system 1695 liters of water was enough. The application efficiency under drip system is about 2.27. In Santa Rosa plum, irrigation at 50% field capacity gave better growth and economic returns.

Pest and Disease Control

In apple and pear about a dozen pests are causing serious damage to the crops. Most important ones are: San Jose Scale, Woolly apple aphid, Root borer, Blossom thrips, Codling moth and European red mite. San Jose Scale can be effectively controlled with eco-friendly miscible spray oils at 2% concentration when applied during February-March. Woolly aphids can be controlled through soil application of Phorate or Carbofuran granules during May and October/November. For codling moth, pheromone trapping was found to be effective, and certain bio-control agents have also been identified in apple orchards for controlling certain insect pests.

European red mite is becoming a very serious pest for apple in H.P. and J&K. The pest attack causes premature leaf fall. Late dormant sprays of miscible oil provide effective control of eggs. The mite can also be controlled by sprays of Dicotol (0.05%) followed by Malathion (0.05%).

. In peach, leaf curl aphid can be controlled by pre-bloom sprays (at pink bud stage) of Dimethoate (0.03%) or Monocrotophos (0.04%). Fruit fly in peach, apricot and plum can be controlled through foliar spray of baits consisting of Malathion (0.1%) + 1% sugar. Stem and shoot borers causing damage to peach, plum, apricot, almond and cherry can be controlled by inserting 0.5g of PDCB (paradichlorobenzene) into the holes and plugging them with mud.

Apple scab caused by Venturia inaequalis is a serious disease causing maximum economic loss. A sound forecasting and early warning system has been developed for prediction of scab attack. Also, a judicious fungicide spray schedule has been devised. Under high disease pressure, systematic fungicides performed better, while under low disease pressure Ergosterol biosynthesis inhibiting (EBI) fungicides were as good as protectants.

Ascosporic inoculum produced by over-wintered apple leaves could be substantially reduced by giving post-harvest applications of Bavistin (0.1%) and EBI chemicals Penconazole (0.5%) and Flusilazole (0.01%) as preharvest fungicidal sprays control scab during storage. Powdery mildew disease can be kept under check by pruning and spraying wettable sulphur (0.2-0.3%) or Karathane (0.05%) during dormancy, bud swell and petal fall stages.

Collar rot and white root rot diseases in apple occur mainly in poorly drained soils. Proper drainage of orchards and soil drenching with 0.1% Carbendazim for white root rot and with 0.3% Mancozeb or 0.3% Ridomil MZ for collar rot are effective. Virus and virus-like diseases such as mosaic, chlorotic leaf spot, rubbery wood and others have been reported. Virus cleaning through tissue culture and supply of virus free bud wood material are being pursued to contain further spread of the viral diseases in apple.

Intercropping

As apple is grown in hill slopes, seasonal intercropping accompanied by working the soil is not encouraged in order to prevent soil erosion. Where slope of the land is more than 10 per cent, various grasses and legumes are grown as permanent covers. The tree basins are kept clean and sod culture in the orchards is practiced in the hills of H.P. The grasses are regularly cut when 7-10 cm high and the residue is left in the orchards for decomposition.

The cultivation of legumes and grasses as intercrop is becoming popular in orchards. The important legumes are red clover (Trifolium pratense), white clover (T. repens) and lucern (Medicago sativa). Among the grasses, orchard grass (Dactylis glomerata), fescues (Festuca arundiceae) and Timothy (Phleum pratense) are common. Orchard grass is grown where moisture is high, while fescues is suitable for dry areas or southern aspects. In apple, sometimes filler trees of peaches or dwarf apple plants are grown.

At the young age of orchard plantations, vegetables like cole crops, potato and tomato are grown to supplement the income. However, this practice is abandoned when the orchard starts bearing after 5-7 years to avoid competition. In Shimla hills and Kullu Valley areas of H.P. cultivation of gladioli for flower and bulb production as intercrop is carried out in the apple orchards since it can tolerate shading.

Harvesting

Apple

In apple, the number of days taken from full bloom to harvest are 132-134 days for Starking Delicious, 138-140 days for Red Delicious and 147-148 days for Golden Delicious varieties in the high hills (above 2000 m) of H.P. In J&K State, the variety Maharaji requires 160 ± 3 days after full bloom for proper harvest maturity. Apple cultivar Granny Smith takes 180 ± 5 days for maturity. The maturity standards for different cultivars of apple are given in Table 1.

Table 1. Maturity Indices for Commercial Cultivars of Apple

Cultivar	Days from full bloom	Firmness (kg)	TSS (%)
Royal Delicious	125 ± 5	8.2 ± 0.40	13.0 - 15.0
Red Gold	122 ± 3	8.3 ± 0.20	12.0 - 13.5
Red Delicious	134 ± 5	8.4 ± 0.40	10.0 - 14.0
McIntosh	135 ± 4	6.8 ± 0.25	11.5 - 13.5
Golden Delicious	148 ± 6	8.4 ± 0.40	12.0 - 14.5
Granny Smith	180 ± 5	8.7 ± 0.30	11.5 - 13.0

Pre-harvest fruit drop in apple could be effectively checked with the application of 10 ppm NAA 20-25 days before harvest. Pre-harvest fruit drop is quite serious in early ripening cultivars like Tydeman's Early, Red Gold and Pippins, where loss of 40-60% crop load is often experienced. Mid-season Delicious group cultivars also experience 15-20% loss.

In apple, fruit color development in warmer and lower (below 1800 m) elevations is generally poor. Application of Ethephon (2-chloroethyl phosphonic acid) about 10 days before harvest improves red coloration. Induction of early maturity by 7-10 days in mid hills situations could be achieved through an application of 500 ppm of ethephon + 10 ppm of NAA 3 weeks before harvesting. In H.P, two pre-harvest sprays of 0.5% CaCl2, followed by a 0.03% surfactant like Tween 20 for 1-2 minutes have been recommended for better shelf-life of apple.

In apple, for obtaining better grade fruits of optimum size a post-bloom spray of 100-200 ppm ethephon 2-3 weeks after full bloom has been recommended both in H.P. and J&K.

Pear

Harvesting time of pear is determined by slight change in fruit color and easy separation of fruits. Fruits should not be allowed to ripen to soft condition on the tree and should be harvested when they are still hard. Depending on cultivars fruit may be ready for harvest from 70 days after full bloom (viz. cv. China pear) to 135 days after full bloom (viz. cv. Barlett). The Bartlett pear normally takes 122 ± 3 days in valley (Kuala) areas of H.P.

Cherry

Sweet cherry should be harvested when fruit stalk separates easily from the spur and taste develops typical to the cultivar. From full bloom it takes 45 days in cultivars like Guigne Noir and Gross Lucenta, whereas in cultivars like Bigarrean Napoleon the number of days varies from 65-75 days.

Peach

For harvesting peach at the right time, the proper color development in fruits and pit browning are considered as reliable guides. In yellow fleshed cultivars, deep orange color development on fruits is associated with proper maturity. At least 5% of the pit area also should be brown if the fruit is to develop good flavor at ripening. In the peach variety July Elberta, it takes about 90-95 days after full bloom, whereas on cv. Elberta it normally takes 100-105 days. In low chilling cultivars like Flordasun, fruits should be harvested at 50% color development stage.

Plum

In free-stone cultivars of plum, pit browning up to 5% of the pit area is a good index of maturity. For proper fruit maturity, early cultivars like Sharp Early and Formosa take 95 to 105 days, mid season variety Santa Rosa requires 115-120 days and late cultivars like Grand Duke take 130-135 days after full bloom.

Apricot

Apricot fruits of a single tree ripen for a period of about 3 weeks. The fruits develop detectable flavor when allowed to ripen on the tree. The early cultivar Charmagz matures within 75 days from full bloom, while the late cultivar Quetta requires 125-130 days for proper harvest maturity.

Yields

There is a wide range of variation in fruit yield in different States and in different cultivars. The average yield of apple and pear have been indicated earlier in Table 2. Although average productivity of apple in H.P. has been shown as 9.4 t/ha, some well managed orchards of Delicious apple in Kotgarh area of the same State yield about 50 t/ha. Similarly, yield record of 35 t/ha has been achieved in 12 year old plantations of colored mutant apple cultivars on MM 106 rootstocks under high density planting system (2222 plants/ha) in a research station in Shimla hills of H.P.

The average yields of different stone fruits have been indicated earlier. In brief, State-wise average yield of different stone fruits are given in Table 2.

Table 2. Average Yield of Different Stone Fruits by State

State	Fruit Yield, T/ha			
	Peach	Plum	Apricot	Cherry
H.P.	0.50	1.31	0.42	-
J&K	0.73	0.74	0.20	1.73
U.P.	2.11	2.50	0.28	-
NEH Region	2.50	1.18	-	-

Fruit and Vegetable Marketing

The biggest wholesale market for apple is the Fruit and Vegetable market at Azadpur, in Delhi. About 70 per cent of the total trade of apple is distributed through this market. The seasons of market arrivals from North Western States are as follows:

— H.P. - July to October with a peak in August - September

— J&K - August to November, with a peak in September - October

— U.P. - June to October, with a peak in July - mid September.

There are a number of marketing channels, of which the predominant are:

a Farmer - Pre-harvest contractor - Commission agent - Wholesaler - Retailer - Consumer.

b Farmer - Forwarding agent - Commission Agent - Wholesaler - Retailer - Consumer.

c Farmer - Commission agent - Wholesaler - Retailer - Consumer.

The sale through pre-harvest contractors is the most important system of marketing. Normally, the small orchardists sell their crop at flowering stage to contractors who organize plant protection practices, picking and packaging of fruits.

The medium and large orchardists prefer to market their produce through channels (b) and (c). In H.P., about 65% of total apple produced is marketed through these two channels. Some Growers' Co-operatives and Government controlled marketing Corporations (like HPMC in H.P.) also get involved in the apple market. For example, in H.P the volume of trading handled by Growers' Co-operatives is about 3.5 to 4.0 per cent, while another 2-2.5 per cent of the total produce is marketed by HPMC.

The National Horticulture Board of the Government of India regularly publishes and announces the wholesale price and market arrival figures of apple in different terminal markets for the benefit of the growers. Some quantities of apple are placed in cold storage facilities.

The Himachal Pradesh Horticultural Produce Marketing and Processing Corporation Ltd. (HPMC) has set up modern fruit packing houses in different parts of the State. Each packing house has a capacity to grade and pack 55,000 to 275,000 cartons of fruits. Cold storage facilities at the production centers and terminal markets at Delhi, Mumbai, Chennai and Calcutta have been created. Similarly, the HPMC has undertaken large scale processing of temperate fruits like apples, peaches and others.

In H.P, grading and packing of fruits have been organised reasonably well. In apple, 7 size grades have been prescribed and dimensions of packing boxes, size of wrapping papers and number of layers of fruits for each grade have been standardized. In apricot, plum and peach, three size grades have been prescribed. Presently Corrugated Fiber Board telescopic tray pack cartons for apple and smaller size universal CFB cartons for stone fruits are in use. These are available with HPMC at their various packing stations. Plastic crates are also used for cold storage and for processing units.

Picking and packaging of fruits are done in different ways at different sites. In U.P hills, trees are generally strip picked into small wicker baskets or small gunny bags which are emptied into larger baskets for carrying to packing stations. Both U.P. and J&K have developed packaging units for different temperate fruits.

India presently exports a small quantity of apple (5.95% of total fresh fruit export), mostly to Bangladesh and Sri Lanka. Apple cultivars like Red and Royal Delicious, Ambri and other new colored cultivars are suitable for the export market. According to the Trade Year statistics of 1993, India exported only 7000 tons of apple during that year.

Processing

The HPMC in Himachal Pradesh owns two fruit processing plants with a combined capacity of 30,000 tons annually. The plants are equipped with modern machinery for sixfold juice concentration and aroma recovery. The plants have also facilities for aseptic bulk packaging and pasteurisation. The HPMC manufactures mainly primary products like fruit pulps, single strength juices and juice concentrates. It covers fruits like apple, pear, peach apricot and plum. The pomace of the fruits after juice extraction is used for cattle feed, extraction of pectin etc.

In J&K, low grade fruits are processed for various products like apple juice, jam, jelly, fruit bars, rings, vinegar etc. from pome and stone fruits. In U.P. hills also, various types of processed products are prepared from temperate fruits. The units are mainly small and medium in size and not to the scale of HPMC. On the research side, important cultivars of apple were screened for their suitability for preparation of juice, jam, dehydrated rings, chops, cubes and preserves.

Golden Delicious cultivar was found to be the multi-use cultivar to meet the FPO quality requirements. A new process for canning of apples called osmo-canning has been developed. Similarly, process for preparation of cider, wine and vermouth have been developed. Utilisation of apple pomace for production of pectin and natural colors have been attempted. Although CA storage concept has not yet become popularised, fruits placed in CA store at Shalimar in J&K have shown that this system of storage has great promise for temperate fruits.

Potential for Fruits

Both North-West and North-Eastern regions of India offer large areas ideally suitable for cultivation of pome and stone fruits. In the North-West India, stone fruits like peach, plum and apricot come up well at elevations between 900 to 1500m with an annual rainfall of 90-100cm. Apple, cherry and pear are commercially successful at elevations between 1500-2700 m above msl. The cold arid regions between 1550-3650m with annual rainfall of 25-40 cm are again highly suitable for dry nuts and drying type of apricots. There are vast tracts of land still available for further expansion of these crops.

In the North-Eastern Hills, excepting the rain shadow belts of Arunachal Pradesh and high altitude Lachung area of Sikkim, apple may not be successful commercially. Other crops like pear, peach, plum and apricot offer good scope for further expansion.

The low productivity and poor quality of apple are often linked with monoculture of a few old cultivars and their degeneration over the years. The U.P hills, particularly Kumaon division, has the unique advantage of harvesting apple fruits for early market. Similarly, the rainshadow belts in North-Eastern Hills can offer good quality apple for the eastern Indian markets, thus reducing the cost of long distance transportation from North-Western Hills. Part of the markets of Bangladesh can be captured by the fruits of the North-East.

There is good scope for introduction of new promising cultivars, replacing the Delicious group. Similarly, use of clonal rootstocks of Malling and Malling-Merton series and even their indexed material 'EMLA' selections will greatly change the productivity and quality of fruits. High density planting with spur type cultivars offer good scope. The identification of low chilling peach, plum and pear cultivars offer good possibilities for their cultivation in the low hills and in sub-tropical plains. Some of the new hybrids, including scab resistant apple cultivars need verification trials on a commercial scale.

Scientific water management and practicing proper training and pruning of trees including introduction of renewal pruning techniques, will make significant impact on increased production even in the existing orchards. Drip irrigation, in-situ water harvesting and correcting macro and micro-nutrient deficiencies will go a long way in bringing notable improvement in productivity as well as fruit quality. By adopting IPM strategy and organic farming practices, selected export markets can be targeted well.

There is immense scope for increasing various processed products of pome and stone fruits, for which technologies are available. The existing capacity of the most organised processing unit of HPMC is only marginal as compared to the volume of fruit available for processing. The HPMC utilised only 1 to 1.5 percent of total cull fruit available. The present combined capacity of two units of HPMC is 30,000 tons annually, which can be easily raised to 50 to 75 thousand tons.

The expansion of grading and packing stations, their further modernisation with mechanical grading equipment, use of CFB boxes and more number of pre-cooling and cold storage units will improve the marketing system and enhance marketability of the produce. Apple has been identified as one of 6 most promising fruits for fresh fruit exports.

Constraints in Fruit Production

Large number of old orchards (more than 30 years old) are showing decline in terms of growth and fruit yield. Such old trees do not produce adequate extension growth. Large scale replanting is therefore needed.

Delicious group of cultivars constitute the major share (about 83% in H.P.) of apple production in the country. These cultivars are self unfruitful and need cross pollination to ensure good fruit set. Interplanting polliniser cultivars in the proportion of 25 to 33 percent is necessary for good fruit set, and choice of wrong pollinisers and their inadequacy in number often result to low productivity.

In many countries, Delicious group has been replaced or is in the process of replacement with more promising cultivars. The need for injecting new blood into the apple industry through spread of new cultivars (spur types, color mutants, strains of Gala, Red Fuji; scab resistant cultivars, bud sport selections of Royal Delicious, and some of the promising hybrids) is urgently felt.

Some of the spur type and colored mutants are already popular with farmers and high density planting has also caught the imagination of developmental departments and agencies both in H.P. and J&K. The research system has already identified Early, Mid and Late cultivars for different agro-climatic regions. The low chilling cultivars of stone fruits have also covered large tracts of the sub-tropical plains of Punjab, U.P and H.P. For the hills, promising cultivars identified need further spread.

Generally, apple is grown in marginal land and fertilisers are not applied according to the requirements of the trees. The water and fertiliser use efficiency is generally poor. Also, spring frost and hailstorms are adverse weather parameters leading to low productivity. Research results have shown that through proper orchard management practices (soil and water conservation and fertiliser application) the fruit yield can be doubled in the existing orchards. The adoption of improved production technology developed by the research system can bring visible and perceptible changes in the temperate fruit industry in India. Technologies like use of clonal rootstocks, introduction of renewal pruning techniques and micro nutrient applications have not been transferred and adopted at a satisfactory level.

Apple scab disease has been the major plant protection problem in apple in J&K and H.P, whilst U.P hills are comparatively free from the disease. Apple scab forecasting system developed and the chemical control schedule prescribed have helped in reducing loss due to apple scab to a considerable extent. Apple growers are adopting the prescribed schedule of chemical sprays to control the disease.

For checking entries of diseased material in the free areas of U.P. and North-Eastern Hills, strict quarantine and selection of elite disease-free mother plants are very essential. Often it is not followed strictly. Some of the virus diseases have also been reported in apple for which biological and serological indexing/detection procedures have been developed. Limited quantity of virus-free budwood is also being supplied. Extreme care is now required to be taken to multiply quality planting material (in apple alone approximately 2 million plants/year) for establishing new plantations.

Most of the orchardists still sell their crop at flowering to contractors as there is no well organised marketing system. Transportation in the hills itself is problematic. Post-harvest management problems originating from poor harvesting (strip picking) and improper packing system (non CFB boxes) and lack of proper pre-cooling and cold storage facilities result in huge (25-30%) loss of fruits.

Capacity of the processing sector is also inadequate. Product diversification, value addition and market infrastructure development would require very substantial investment. The existing processing units are quite old and they require modernisation for which substantial investment is required. CA storage trials have shown good promise. Its extension in larger growing areas is needed. Technology for storage of apple is now known, as a result of which apple is now available throughout the year.

Policies and Plans

Research on pome and stone fruits is conducted mainly by three State Agricultural Universities, namely:

a) Sher-e-Kashmir University for Agriculture Science & Technology, J&K;

b) Y.S. Parmar University for Horticulture and Forestry, H.P.; and

c) G.B. Pant University for Agriculture & Technology, U.P.

A good number of research stations of these Universities located in major pome and stone fruit growing belts are engaged in temperate fruit research. The Indian Council of Agricultural Research (ICAR) through All India Coordinated Research Projects on Fruits and Post-harvest Technology and another Network project on Apple Scab disease has further strengthened the research activities in deciduous fruits.

A few long established temperate fruit research stations namely at Shalimar Bagh in J&K, at Mashoobra in H.P. and at Chaubatia in U.P. hills have made commendable progress in temperate fruits. Both State Governments and the ICAR provide financial support to strengthen fruit research in these Universities. A large number of ad-hoc research projects on pome and stone fruits funded by ICAR are also generating good information on these crops.

During the 8th Government Developmental Plan the ICAR established a Central Institute for Temperate Horticulture (CITH) with its headquarters at Srinagar in the J&K State, with a regional station at Mukteshwar in the U.P. Hills. Both these research stations will work exclusively on temperate fruit crops. This new Institute will receive major support during the 9th Plan period. In the North-Eastern Hills region, the ICAR Research Complex for NEH Region with its headquarters in Meghalaya State and regional units in each of the 5 other States are engaged in fruit research.

On the developmental side the State governments are engaged in nursery production of quality planting material. For example, in H.P. alone currently there are 600 nurseries in private and public sector producing and distributing more than 0.8 million plants of apple alone, every year. There are ambitious programmes in all the States to further expand/replant with new improved cultivars. Apple scab disease control and post-harvest processing sectors are getting focused attention in Government developmental plans.

The Directorate of Marketing and Inspection of the Government of India has framed grade standards for apple, plum and William pear. The organisations like National Horticultural Board (NHB), National Co-operative Development Corporation (NCDC), Agricultural and Processed Food Products Export Development Authority (APEDA) etc., are providing incentives to traders and exporters to improve their infrastructural facilities like grading and packaging centers, refrigerated transport, setting up of pre-cooling, cold storage, auction platforms etc. The NCDC is undertaking procurement and marketing of apple on a limited scale. The NHB has set up a market information service for the benefit of growers.

Fruit Production in China

China is one of the earliest and most important centers of origin of cultivated plants in the world. Many deciduous fruits such as peach, Asian pear, apricot, plum, jujube, chestnut and filbert that are grown today are native to China. From the fossil remains of filbert and chestnut dug up from Banpo village, Xian, Shaanxi province and peach stones discovered in archeological investigations in Wu County in Jiansu province, it is evident that pomology is an agricultural science that started in China nearly 6000 years ago.

Manuscripts mentioning 17 different fruits, mostly deciduous, can be found in Shi Jing, an ancient record of songs written about 1000 BC. Cultivar differences of winter peach and stoneless jujube were recognised by Chinese ancestors at least as early as the early part of second century B.C. Due to this long history of fruit cultivation, an abundance of fruit growing experience and production practices has been accumulated in cultivar selection, grafting, fruit thinning, pest control and storage technologies. A great number of valuable varieties were developed, some of which are recognised even today in fruit production programmes.

After the establishment of the People's Republic of China in 1949, the fruit industry passed through its recovery and stable development stage which led to a period of rapid development. In 1997, the total production of fruits in China was 50.89 million tons, out of which 17.22 million tons were apples and 6.42 million tons were pears. They accounted for 11.2, 31.8 and 44.4 percent of the world production respectively. Fruit production in total or individually for apple and pear assumed the first place in the world.

The fruit industry ranks third in the country after grains and vegetables and exceeds that of cotton, forestry, oil and sugarcane. Per capita availability of fruits per annum increased 6-fold from 6.8 kg in 1978 to 41.2 kg in 1997. It has been the policy of the Government to encourage the development of the fruit industry in the mountains and hilly areas, leaving the fertile lowlands for the production of grains and cotton. Under this policy the loess plateau in the Northwest, highlands of the Southwest, other hilly regions and mountainous areas have become bases of commercial fruit production.

Development of fruit growing on the highlands has several advantages. Firstly, due to more exposure to sun light and greater variation in daily temperatures, fruit trees growing in the highlands are likely to produce quality fruit with good color and higher sugar content. Secondly, by growing fruits the farmers can obtain much higher returns than from field crops. An added benefit of cropping the highlands is the improvement of socio-economic conditions of undeveloped highland communities.

Furthermore, as commercial fruit orchards got established in these areas, contour terraces, water conservation projects, roads and storage facilities were established. Other services related to fruit production were also improved. These, accelerated the economic development of the rural highland areas and the fruit industry was often referred to as a green revolution with fruit trees acting as environment cleaning agents.

Deciduous Fruit Cultivation

Among deciduous fruit crops grown in China, apple is far ahead in acreage and production, while pears, peach and grapes are also leading crops. Apricot, plum, mume, jujube, persimmon, walnut, chestnut and kiwifruit are next in importance. Other deciduous fruits such as cherry, almond, pomegranate, fig, filbert and ginkgo are also cultivated but to a limited extent. Hawthorn, Roxburgh Rose and Seabuckthorn are new fruit crops that have been recently domesticated. In China, a large portion of the deciduous fruits are produced in the North along the Yellow river valley. However, the Yangtse river basin areas are also important for the production of sand pear, honey peach and mume.

The apple crop is grown on 32.66 percent of the total area under fruit cultivation and it contributes 33.8 percent to the total fruit production. This makes apple the most important fruit crop among deciduous fruits in China. There has been a 4.2-fold increase in acreage and 7.5-fold increase in production of this crop during the last two decades. The major fruit producing areas are along the Bohai Sea, the Yellow river lowlands, Northern foot of the Qing-ling mountains, the Loess plateau in the Northwest and the highlands of the Southwest.

The major producing provinces are Shangdong, Lianing, Hebei, Henan, Shaanxii and Shanxi, while minor production is undertaken in Gansu Xijiang and Jiangsu. During the 1980's more than 20 cultivars were under cultivation. Nearly 60 percent of the apples produced were of cultivars such as Ralls Janet, Golden Delicious, Red Delicious and Jonathan. At the present time, however, over 70 percent of plantings are of Red Fuji, Starkrimson, Jonagold, Golden Delicious and Gala apple cultivars.

The second major crop among deciduous fruits is pear, which covers an area of 924,000 ha with a production of 6,415 million tons a year. Its share in total deciduous fruit crop area is 10.6 percent and production is 12.6 percent. Cultivated pear in China is classified into four groups. White pear (Pyrus bretschneiderie) is mainly grown in Northern China, Hebei, Shandong, Liaoning provinces and accounts for 60 percent of pear production in the country. A great number of cultivars exist in this group; some of these cultivars, notably Ya-Li, Xue-hua-li, Lai-yang-ci-li and Dong-guo-li produce fruits of excellent quality with crispy, juicy and sweet flesh and relatively few stone cells. Sand Pears (P. pyrifolia) grow almost wild in the Yangtze river valley.

They adapt well to wet and high summer temperatures and the popular cultivars Cang-xi-li and Bao-zhu-li are locally important cultivars. Huang-hai-li and Jin-shu 2 are newly developed cultivars and are extensively used in new plantings. The Japanese cultivars also belong to the sand pear category. Some Japanese cultivars such as New century, Kosui and Shin sui are also important cultivars in this area. Fruits of white pear and sand pear are juicy, crisp and sweet. They do not require a ripening after harvest and could be used as a dessert fruit immediately after picking.

The Ussurian pear (P. ussuriensis) is the most hardy of all Pyrus species and is grown in the areas North of the Great Wall, especially in Northeast China. In general, fruit quality of the cultivars derived from this species are very much inferior to those of the white pear and sand pear. The fruits are usually smaller and require a period of post-harvest ripening to become edible. After proper ripening they become soft with a strong aroma and acceptable quality. Representative cultivars of this group are An-li, Da-xiang-shui-li, Nan-guo-li and Jing-bei-li. European pear is not commonly found in China and it is only grown to a limited extent in a few localities.

Peach is native to China. There are two major groups of cultivars and ecotypes. The Mitao cultivars are generally cultivated in North China. They tend to bear larger fruits with firmer flesh. The principal producing areas are located in Feicheng and Yidou in Shandong province, Shenzhou of Hebei province, Liquan and Fuping of Shaanxi province, Zhenging and Zhangye in Gansu province. Leading cultivars include Feicheng Tao, Shenzhou Mitao and Yidou Mitao. Shumitao peach, also known as Honey peach, is mostly grown in Southern parts of China and it is well adapted to wet and high summer temperatures in the Yangtze river basin.

Fruit of this peach is tender-fleshed, melting and juicy with high sugar content. The important areas of peach production are close to the cities of Nanjing, Wuxi, Shanghai, Hangzhou, Ninbo and Fenghuo. Fenghua yulu and Baihua shuimi are the best known cultivars in these production areas.

Peach fruits grown for fresh markets in China are almost all characterized by white flesh, melting and juicy clingstone types, while non-melting and yellow-fleshed types are mainly used for canning. These canning types thrived best during the period between 1970 to 1980's, but decreased in the 1990's. On the other hand, early peach and nectarine cultivars and their area of production have increased recently.

Plum (Prunus salicina), apricot, mume, Chinese cherry (Prunus pseudocerasus), and persimmon are all native to China. They are grown almost all over the country, except mume which is restricted to the South China. Major producing areas are also located in Hebei, Henan, Shandong, Shanxi and Shaanxi provinces. Oriental plums differ from the European plum and have two categories of cultivars, red colored and yellow skin types. The red types possess either yellow or red colored skins, but the yellow types have exclusively yellow flesh only. There are more than 800 cultivars under cultivation, but Zu-li in Zhejiang, Fu-rong-li and Mi-liin Fujiang are highly preferred by consumers.

Apricots are found mostly in North China, mainly concentrated in the Yellow river valley. The many cultivars under production have different uses. Some are grown for the fresh market (the larger fruited types reaching up to 200 gm in size). Although the Yangtze river valley is not the main producing area for apricots, many important local cultivars exist. Some bear high quality fruits of large size and they are well adapted to humid weather prevalent in the area.

Both European cherry and Chinese cherry are grown in China. Chinese cherry (Prunus pseudocerasus) has been grown in China for over 2000 years, and bears more fruits than the European cherry. Chinese cherries also have a wider range of distribution. Taihe in Anhui, Zhuji in

Zhejiang, Zhengzhou in Henan, Qingdao, Zaozhuang in Shandong and Lantian in Shaanxi province are the major producing areas. The European cherry cultivars were introduced to China in 1870. Their production is limited to Yantai in Shandong, Luda in Liaoning, and Beidaihe and Changli in Hebei province. Napoleon, Black Tartarian and Bing are major European cultivars grown in China.

About 860 cultivars of persimmon exist in China. They bear fruit varying greatly in shape, size, color and ripening period. Most of the cultivars are astringent in taste at harvest time. Persimmons originated in Southern China, but the largest production is confined to the Yellow river valley. Annual production of persimmon is around one million tons and Shaanxi, Shanxi, Hebei, Henan and Shandong provinces account for 70-80 percent of the production. Fuyu and Jiro are major cultivars of sweet persimmon from Japan.

Mume (Prunus mume Sieb et Zucc.) has been cultivated in China for over 3000 years. The wild forms can still be found in mountainous areas between Hubei and Sichuan provinces. Since they bloom in early spring and are sensitive to frost damage, the production areas are limited to Yangtze river valley and its Southern parts. Zhejiang, Jiangsu and Hunan provinces are the major producers of mume fruit.

Management of Orchards

Methods in training and pruning of trees vary depending on climatic and soil conditions, fruit species, cultivars, rootstock, spacing and management systems used in orchards. Two basic training systems are, however, commonly adopted in the deciduous orchards in China. They are the modified central leader system for apples and pears, and the open center system for peach trees.

Pruning is mainly carried out in the dormant season, but summer pruning is also practiced, especially in peach orchards. Hand pruning is considered a special art and a skill very much needed in proper tree management. Despite the differences in tree forms or the time of pruning, much concerns are centered around the balance between vegetative and reproductive growth.

Fruit growers in China traditionally use a high amount of organic matter as basic source of nutrients for the trees in the dormant season. Animal manures are used extensively and sometimes oil-pressed cakes and chemical fertilisers are also added. These basic nutrient sources are considered to be important for quality fruit production since they not only provide nutrients for tree growth, but also improve soil structure which promotes better root development. Their application is usually recommended during the dormant season up to autumn, after fruit harvest but usually before leaf fall. This application coincides with the third peak of annual root growth.

The fertilisers help trees to recover from the depletion caused by a heavy fruit load and also enhance development of quality fruit buds. Top dressing applications are usually chemical fertilisers which are applied two to four times before bud break, prior to or after blooming and at fruit enlargement stage depending on the requirements of each type of fruit tree.

Since most orchards in China are established in hilly areas, the soils may be low in nutrients and not deep enough for good root development. Therefore, in addition to reclamation work before orchard establishment, soil improvement after planting is also required. As most soils in such areas are around 50 cm in depth (usually with a hardpan), deep tillage is required.

Subsoil tillage is carried out either by enlarging the planting pit up to the tree canopy line or by constructing ditches between tree rows. After removing the subsoil, pits and ditches are filled with a mixture of top soil and organic manures such as farmyard manure, straw, compost etc. If orchard soils are clayey, 1.5-2.25 tons of lime hydrate is recommended for incorporation during subsoil tillage to help regulate soil pH. In young orchards, deep tillage can be carried out in spring, summer or autumn, while in older orchards this practice is carried out after harvesting the crop.

Since 1980's, the use of straw or plastic film has become popular in orchards established in hilly areas. If straw mulch is used, the soil is initially tilled after incorporating farm manures and a 10-20 cm layer of straw is used for mulching every year. If clear plastic film is used to mulch under tree canopies, a 10 cm soil layer is used to cover the plastic mulch, especially when temperatures go up to 20-25 °C.

Although annual rainfall in Northern China is adequate to meet the requirements of fruit tree orchards, the distribution is erratic and most of the rain is received from July to September. Supplementary irrigation is therefore essential during the period from October to June. In Southern China, irrigation is also required during the hot and dry spells in July-August. Surface irrigation systems are mostly used in China, while drip systems are preferred in saline soils. Sprinkler irrigation is also being installed in several areas in recent times.

Deciduous Fruit Crop Development

The fruit industry in China has reached a new stage of development with a massive increase in growing area and production. At the same time, however, this rapid development has brought about many problems and the fruit industry is currently facing many challenges.

Firstly, with such a rapid increase in production it has brought in its wake deficiencies in post-harvest handling, storage facilities, processing and marketing which are unable to cope with the volume of fruit produced. As a large quantity of fruits reached markets, disposal within a short period could not be achieved, resulting in depressed prices and considerable losses which even forced growers to abandon harvesting the crops.

Secondly, due to the rapid expansion in production area, the use of low quality planting material was sometimes difficult to avoid and some of the cultivars grown by farmers are of very poor fruit quality. Another factor was the lack of training given to fruit growers, especially in the new fruit growing areas. Poor management has also resulted in lower yields per unit area in addition to the use of inferior germplasm. This situation will reach catastrophic proportions in the near future when about 40 percent of the young plantings reach bearing age.

It was found that about 66 percent of the total fruit production was contributed by apple, pear and citrus. Many of the cultivars of these fruits have a short harvest season which contributed to unmanageable surpluses. Quality was another factor since nearly 20 percent of fruit produced was of smaller size and unacceptable quality. Looking into these problems, it was also revealed that only about 30 percent of fruits produced were of good quality while the remaining 50 percent reached average quality standards. It was also found that only about 1 percent of the harvest received post-harvest treatments such as fruit cleaning, waxing, grading and packing.

Currently available storage facilities can accommodate only 15 percent of production, leaving 85 percent of the harvest to be disposed within a short period. This has resulted in severe post-harvest losses and lowering of prices bringing about economic ruin to farmers. Lack of storage facilities has resulted in about 20-25 percent of the total fruit produced perishing after harvest.

Under these circumstances, new Government policies for the fruit industry are directed towards giving high priority to fruit quality, yield per unit area and better economic returns to fruit producers. Whilst total extent under fruits will remain at the same level, there will be changes in the combination of species and cultivars used. Regulations will be introduced to effect these changes.

In order to maintain the proportion of the area under citrus, apple and pear in a ratio of 6:4 with other fruit crops, part of the hectarage of these crops that are grown under sub-optimal agro-ecological situations will be replaced by other fruit crops. It has also been planned to replant about 1.6 million ha of old orchards and about 3 million ha of plantings with low quality or low yield potential will be rejuvenated by topworking using newly developed superior cultivars.

Cultivar regulation will also include the spreading of the harvest season. A proper mix of early, mid and late cultivars of all fruit crops will be promoted in a ratio of 10, 30 and 60 percent respectively. In apples, the cultivar Anna will be grown for early season harvest, Tsugaru, Gala, Starkrimson and Golden Delicious for mid-season and Red Fuji and Jonagold for late season production.

The current policies will also ensure fruit quality and better economic returns rather than increases in extents or volume of production as earlier encouraged in the 1980's. One of the major goals of the new policies include the increase in the percentage of quality fruit from the present 30 percent to 40 percent by the year 2000, and a further increase up to 60 percent by the year 2005.

Deciduous Fruit Production in Nepal

Nepal is a predominantly agricultural country. Agriculture is the lead sector for the national economy and accounts for about 42% of the GDP. About 81.1% of the population or about 3.3 million families are engaged in agriculture. Land holding per family of 5-6 members is 6.5 ha in the hills and 1.8 ha in the terai.

The Nepalese hill economy is characterised by typical subsistence agriculture based on cereal crops, and is practiced on terraces of often very steep slopes which are subjected to a great loss of top soil by erosion during heavy rain. Hilly soil is generally acidic and with poor nutrient content, especially nitrogen.

Administratively, the country is divided into 75 districts and five development regions, namely, Eastern (16 districts), Central (19 districts), Western (16 districts), Mid-Western (15 districts) and Far-Western (9 districts).

Agro-ecological Classification

The topography of Nepal is extremely variable ranging from 60-300 m above sea level (m a.s.l) in the southern plains to 8848 m.a.s.l. in the north, which is the highest point on earth (Mt. Everest). Big variation in altitude occurs within short distances due to which it enjoys all types of climates. Physiographically, the whole country can be divided into five zones.

Commercial fruit growing, especially the deciduous type is not traditional in Nepal. Therefore, it is limited largely to homestead gardens and fruits are used mainly for domestic consumption. The local peach, pear, plum and walnut are grown in scattered locations, they are of inferior quality and have very little commercial value. This however, indicates the potential for growing temperate fruits in Nepal.

For the first time in Nepal a Department of Agriculture was established in 1925. After this, a number of different fruit species, mostly temperate fruits were introduced to Balaju and Godavari orchards in Kathmandu. Only after 1950 several promotional activities on fruit development were undertaken in Nepal. For the first time a Horticulture Development Section was established under the Department of Agriculture in 1955.

Several improved cultivars of Asian pear, peach, plum, persimmon, cherry and apple were introduced to Singh Durbar and Kakani farms and cultivar performance studies and propagation activities were started with bilateral assistance from USAID until 1963. From this time the Government of Nepal placed special emphasis on fruit development in the hills. The period from 1960 to 1973 was a very crucial period because this is the time when six temperate horticultural stations were established at different locations of the country with support from the Indian Cooperation Mission (ICM).

Many new cultivars of deciduous fruits were introduced into these stations from India. In these stations a number of activities were started including cultivar performance studies, planting material production and distribution, training of farmers etc. As a result, the area under deciduous fruits increased dramatically. A number of new temperate fruits and additional new cultivars were introduced and key horticultural stations were strengthened during 1977 to 1980 under the Hill Agriculture Development Project assisted by FAO.

Until the early seventies, the requirements of planting material of these fruits were met mostly by imports from

India but after the mid-seventies several private nurseries were established in the major growing areas and self-sufficiency was almost achieved. No significant research work on fruits in general, and deciduous fruits in particular, was carried out in Nepal in the past. Only preliminary studies on cultivar evaluation, propagation and insect pest management were undertaken.

Present Situation of Fruit Production

At present, deciduous fruits are considered as the most important fruit crops of Nepal. These fruits are grown successfully in mid and high mountainous areas from the Eastern to the far Western zone of the country. Sub-humid and dry temperate areas in the inter Himalaya region where the elevation ranges from 1800-2800 m.a.s.l. are considered to be most suitable for particularly high quality apple production. Such rainshadow or low rainfall areas are located in the Western and mid Western mountainous regions.

On the other hand, the humid temperate regions where the rainfall is high and are fairly wet throughout the growing period are suitable for the cultivation of other deciduous fruits. Some low chilling (<1000 hrs) apple cultivars are being grown at low altitudes, as low as 1200 m.a.s.l.

Important Cultivars in Nepal

Several cultivars of different kinds of deciduous fruit species were introduced into Nepal at different times and for several programmes. These cultivars were imported mainly from India, the United Kingdom, Italy, Israel, Japan and USA. They were planted in different horticultural stations and suitable cultivars were propagated and distributed to the farmers. The important cultivars of different deciduous fruits available in Nepal are given below.

Apple

Both high chilling and low chilling cultivars of apple are cultivated in Nepal. The principal high chilling cultivars are Red, Royal and Golden Delicious, Mc Intosh, Jonathan, Rome Beauty, Granny Smith, Richared, Golden Spur, etc. Among all these cultivars the Delicious group covers a major area as their fruit quality is excellent. The mid chilling cultivars are Katza, Red June, Cox Orange Pippin, Crispin and Summer Pippin. The low chilling cultivars are Anna, Vered, Tropical Beauty, Winter Banana etc.

Rootstocks: Crab apple (M. baccata) seeds are used extensively for raising rootstocks for apple in Nepal. Edy Mayal which is commonly found in the wild is also being used on a limited scale as a rootstock. Clonal rootstocks such as M9, M26, M27 of the Malling series and MM101, MM106, MM111 of Malling Merton series have been introduced to a few horticultural stations and are used for propagation on a very limited scale.

Production Areas: High and mid chilling cultivars are mostly grown in an altitude range of 1800 - 2800 m.a.s.l, where chilling is more than 1000 hours; low chilling types are cultivated at elevations as low as 1200 m.a.s.l. and where chilling is 600-1000 hours. As far as elevation is concerned, apple can be grown throughout mid and high mountain areas from Eastern to the far Western region; however, due to high humidity and heavy rainfall during the growing period the most suitable areas for quality apple production are confined to the mid and far Western region where dry to semi humid conditions exist.

Pear

There are two kinds of pears grown in Nepal. The oriental pear or sand pear locally known as Naspati or Pharping Naspati (Pyrus pyrifolia) is very popular as it needs low chilling hours (<1000); it is very well adapted to the warmer temperate region of Nepal.

Japanese cultivars including Shinsui, Shinko, Kosui, Hosui, Chojuro and Okusankichi were introduced into Nepal about 10 years ago under a Japanese Aid Horticulture project. These cultivars are successful under warm temperate conditions and fruit quality is excellent. These cultivars, however, require very careful management.

The European pears (P. communis) were introduced into Nepal many years ago. The better adapted popular cultivars of European pear are Bartlett, Quince, Anjou and Conference, which are also usually grown in cool temperate regions.

Rootstocks: A wild variety locally known as Mayal (P. pashia) is extensively used as a rootstock. Seedlings are raised from seed and grafting is done on them.

Production Areas: The pear is a very important deciduous fruit crop in Nepal. It is cultivated in both mid and high mountain areas of the whole country.

Peach

So far, a total of 44 improved cultivars of peach have been introduced into Nepal from many countries. High chill cultivars are Peregrine, Triumph, Elberta, Baby gold, Suncrest, Rhodes, Red Haven, Florida Red, etc. Low chill cultivars include Orion, Spring time, French Early, Cardinal, Armgold, Florida, Kuratake Wase, Texas, Spring time, Early Red etc. These are all early ripening cultivars, while J.H. Hale and Late Elberta are popular late peach cultivars in Nepal.

Rootstocks: A wild cultivar of peach which is widely found in Nepal is extensively used as a rootstock for cultivated peach, nectarine and almond.

Production Areas: In Nepal, peach is a very common stone fruit extensively grown throughout the mid elevations between 1000-2800 m a.s.l. High chill cultivars are performing well in the high mountain areas, whereas low and mid chill cultivars perform very well in mid mountain areas.

Plum

Two types of plum are commonly grown in Nepal. European cultivars grown are Green Gaga, Early Transparent Gage, Stanley, Ruth Gestetner etc. Japanese cultivars are Santa Rosa, Methley, Satsuma, Formosa, Mariposa, Burbank, Kelsey, Oishi Wase and Shiro.

Rootstocks: A wild plum or peach are commonly used as rootstocks for all plums. Clonal rootstocks of Myrobolan are also used on a very limited scale.

Production Areas: Plums are being successfully grown in areas where peaches are grown except those areas where there is early spring frost.

Apricot

About 13 cultivars of apricot were introduced into Nepal at different times but unfortunately only a few of them have been successful under Nepalese conditions. These cultivars included Blenheim, Titon, Bulida, Rcale Dimola, Prete, Canino, Kaisa, Charmagz, Shakarpara etc. Shakarpara is the most successful cultivar in Nepal.

Nectarine

Popular nectarine cultivars are Independent, Panamint, Arm King, Ruby Gold, Fantasy, Nectaret-2 and Neyorkert.

Persimmon

Popular cultivars are Euyu, Ziro, Zenjimoru, Hiratanonashi, Hachiya and some local selections.

Rootstocks: Local cultivars.

Production Areas: Warm temperate areas throughout the country.

Cherry

Cultivars grown are Napoleon, Satomishiki, Tokasago, Victoria and Bigara. However, none of these cultivars have shown good performance in Nepal.

Planting Material

Nepal was importing deciduous fruit plants from India until the mid 80's. Due to the encouragement from the Government several private nurseries have been established in different regions, and now the country is self-sufficient in the production of planting material of these crops. Due to the expansion of the area under these crops, the demand for planting material is increasing rapidly.

Orchard Management

In Nepal, more than 80% of cultivated land in high and mid mountain regions is on sloping mountainous terrain. In this hilly terrain the contour system of planting is followed. All trees are planted on the contour. The distance between the rows depends on the slope, being closer on steeper slopes and wider spaced otherwise.

On very steep land, terraces are made and the slope of the terrace is kept inwards in order to prevent soil erosion. In such terrain, hexagonal or rectangular system of planting is practiced. Pits of approximately 1 x 1 x 1 m size are dug during the period of September-October and filled with a mixture of compost and soil. Planting of deciduous fruit trees is commonly done through December to March depending upon the altitudes. Planting distance for all these fruit trees is 6 X 6 m (300 plants per hectare) for most parts of Nepal.

Proper training of young plants and pruning of older trees are not strictly followed by farmers in many areas of Nepal. Farmers' negligence of these very important operations has created a big problem in quality production. However, the most common system of training of temperate fruits practiced in the country is the modified leader system followed by the open center system. Pruning is generally done during December to January.

Soil fertility in the mid mountain regions is extremely low, and crops are mainly dependent upon farm-yard manures and compost, and fruit crops hardly receive fertilizers. The three-fourth dose of nitrogen and full dose of compost, phosphorous and potash are applied prior to the dormancy break. The remaining one fourth of N is given soon after the fruit harvest. In Nepal, fruit cultivation is done on marginal uplands where irrigation facilities rarely exist; therefore, orchards are entirely dependent on monsoon rain which occurs mainly during May to August.

Mulching around the trees has been highly recommended to fruit growers, but in practice (except in a few cases) it is rarely followed. Hand weeding is the common practice in Nepal. Shallow cultivation with hand tools at the time of inter-cropping is the only way of working the soil in orchards.

In Nepal, reports on plant protection research work on fruit crops are very scanty. The problem of major pests and diseases of these crops is alarming.

Farmers of Nepal do intensive intercropping so long as there is sufficient space between trees. Farmers usually do shallow cultivation on the surface away from the periphery of the trees. Most of the farmers use maize, pulses, mustard, vegetables, potato and spices as inter-crops.

Harvesting of deciduous fruits is entirely done by hand. Fruits are picked by hand (sometimes beating with stick) and placed in large bamboo baskets. Fruits are collected into one place until they are sent to the market. Actually, no grading is done here except sorting out badly damaged fruits as a result of bad harvesting or due to pest and disease damage.

Marketing Channels

Most of the deciduous fruits are cultivated in remote and isolated mountain areas. The majority of these areas neither have access to motorable roads and marketing channels nor any storage and processing facilities. Fruits are packed in simple bamboo baskets, each weighing about 30 kg, or sometimes in jute bags and transported by porters or mules to nearby markets or road collection points.

In some places it takes several days to reach these points. Once these fruits reach road heads, they are transported by trucks, buses and even airplanes to large markets. After arriving at the markets they are sold to wholesalers or middlemen. Purchasing is done on weight basis or as whole basket load.

Perishable fruits like peach and plum have great difficulty to find markets in remote areas, and farmers are obliged to sell them at very nominal prices; sometimes they may not get buyers at all. From the most remote areas of the high mountain regions, fruits, especially apples are airlifted by plane in limited quantity to large markets. This is the common mode of transport in Mustang and Jumla areas.

In these areas, most of the fruits are sold locally to incoming tourists or to local processors at surprisingly low prices. In these high mountain areas a local method of cellar storage has been developed to store apples. Pre-treated fruits with 0.25% potassium permanganate solution are packed in wooden crates and stacked in rock-lined dark cellars. The humidity in such cellars is controlled by providing running water channels on the floor. In these cellars, fruit can be stored for about 3-7 months depending upon the elevation of the place.

Nepal imports fresh as well as dried and processed fruit products, mainly from India, in large quantities. Among the fresh fruit, apple is imported in bulk. A very small quantity of these fruits is exported to India. It is really difficult to get actual data on imports and exports of fruits. Total fresh fruit imported into Nepal in 1996 is reported to be worth about 100.8 million Nepalese Rupees (about US$2 million) whereas only 0.6 million rupees (US$ 10,000) worth of fruit was exported. Nepal is heavily dependent on imported fruits.

Fruit and Vegetable Processing

There are about 2000 agro-processing enterprises in Nepal, but only about 150 are fruit and vegetable processing enterprises. Processing some deciduous fruits into alcoholic products has become very popular, especially in remote areas. There are about 130 small processing units in the hill districts, out of which 25 are wineries. The second important processed product of these fruits is jam, locally very popular, and is sold specially to tourists, restaurants and hotels. Air dried apple slices also are becoming quite common. Apple fruits that are not sold are sliced and sun-dried locally in the growing areas.

Deciduous Fruit Production Development

Agriculture is the backbone of the Nepalese economy and horticulture is recognised as one of the important sub-sectors of agriculture. It contributes about 13% to the total agriculture GDP (AGDP) but covers only 4.4% of the total cropped area.

Due to the wide range of climatic conditions Nepal has a unique opportunity to develop many fruit crops. During the last two decades, Nepal has been under the process of rapid urbanisation. Demand for fruits is also growing at a fast rate due to the awareness of balanced diet and changing dietary habits of the people. It is quite evident that fruit farming is far more profitable than cereals, where marketing and other essential facilities exist. Per capita fruit consumption is far below the recommended level and only 45% of the local demand of fruit is met by domestic production, while the rest is imported from India.

However, the Agriculture Perspective Plan quotes that 85% of the total consumption is imported. Soil erosion from the intensive cultivation of land and environmental degradation from the rapid deforestation are detrimental to the environment and agriculture sustainability. Soil fertility decline is apparent throughout the hill region of Nepal. Present soil management techniques adopted are unable to meet the overall production demands of the local population.

Nearly three-quarters of the population depend on agriculture for employment. There is no doubt that fruit farming is moderately employment intensive and contributes considerably to farmer incomes. Ecological niches suitable for commercial cultivation of deciduous fruits have already been identified. These areas are becoming more and more accessible every year because of the construction of new motorable roads.

Based on the successful apple production in some pocket areas, the Agricultural Perspective Plan has identified apple as one of the high value priority commodities in the high hills. Accordingly, the Government has given high priority for developing apple production in suitable pocket areas where transport and market infrastructure have already been developed or are in the process of development. Additional areas will be covered by apple orchards under the ongoing Ninth Five Year Plan period.

Constraints in Fruit Production

Almost all types of deciduous fruits are being grown in various parts of Nepal but it has not been demonstrated as an economic enterprise. This is because they are constrained by a number of factors which can be broadly categorised as

a) Infrastructural,

b) Physical and Environmental,

c) Agronomical,

d) Technical and

e) Socio-economical.

Deciduous fruit growing areas are situated in remote places in the mid and high mountains. These areas are not yet accessible to motorable roads and large markets, so farmers are facing big problems in selling their produce which discourages them from venturing into commercial production enterprises.

Essential inputs such as fertilisers, plant protection chemicals and horticulture tools etc. are not regularly available and on time, in these areas. Farmers have to depend upon compost and farmyard manures which are not readily available to fruit crops. As a result, fruit trees are nutrient deficient, untrained, unpruned and unprotected against diseases and pests. These are some of the causes for low yield and poor quality of these fruits.

Fruit cultivation requires large initial investment which the majority of Nepalese farmers cannot afford to make and have no access to any credit. Organised marketing channels, transportation, storage and processing facilities have not yet been developed in all deciduous fruit growing areas of the country. These are very important limiting factors that hamper profitable fruit farming in the country.

Physically, Nepal is situated in a very difficult region of the world. More than 80% of its geographical area comes under hilly and difficult terrain with very steep slopes, loose soil and rocky structure. It is understandable how difficult it is to establish the basic development infrastructures in these areas for a very poor country like Nepal. It is also being threatened by the alarming situation with regard to environmental degradation and soil erosion that is being created by the rapidly growing population, compounded by unemployment and poverty (> 40% are below absolute poverty line).

It might appear that there would be great opportunities to grow a wide range of fruit species at specific locations in Nepal when one considers the enormous range of climatic conditions. However, certain unfavourable conditions also exist in Nepal such as heavy rain during monsoon period, uneven distribution of rainfall, high wind and hailstorms, spring frost etc. More than 90% of rainfall occurs from June through September, precisely when deciduous fruits are developing and maturing. It is difficult to control diseases and insects during such warm humid summers.

In the dry spring months from March to May when temperatures are high with little rain and no irrigation, fruit trees are seriously stressed due to which productivity is reduced. In most of the deciduous fruit growing areas, spring hailstorms are very common, which cause tremendous damage to developing fruits. Many potential deciduous fruit growing pockets in high mountains experience spring frosts that can destroy the flowers at full bloom stage. This spring frost causes more damage to warm temperate fruits like pear, peach and plum. Some important fruit growing areas like Mustang and Jumla experience strong wind in the spring that causes severe flower and fruit drop.

Lack of quality planting materials, mineral fertilisers, and incidence of economically important insect pests and diseases are the most important agronomic constraints to fruit cultivation in Nepal. It is estimated that more than 270,000 deciduous fruit plants are needed annually. The majority of the nursery plants produced in the country are of inferior quality and not true to type. Most of the private nurseries do not maintain quality mother stock and standard rootstocks. Plants produced in the private nurseries are generally of substandard and not healthy.

Another important constraint to fruit production in this country is the incidence of economically important insect pests and diseases. For example, woolly aphis, San Jose Scale of apple, and defoliating beetles of pear and peach, and diseases such as apple scab, pink disease of apple, pear decline, peach leaf curl and foot rot are commonly found.

Technical support constraints are summarised as follows:

— Low priority for fruit research

— Poor allocation of fiscal resources

— Lack of trained manpower

— Weak extension system in horticulture

— Inadequate technical information

The Nepal Agriculture Research Council (NARC) is the sole agent responsible for fruit research. Only a little research has been done in fruit crops to solve technical problems which the growers are facing. There is also a total lack of coordination between research and development.

Every district agriculture extension office has horticulture extension staff but they are completely deprived of technical backstopping. The number of deciduous fruit specialists is very limited in the country. Horticulturists working in extension have very limited access to information on new horticulture technologies and they are not exposed to any specialised training.

Socio-Economic constraints are summarised as follows:

— Lack of experience in fruit growing

— Socio-cultural

— Food habits

— Land tenure

Unlike cereal crops, farmers of Nepal are not experienced in growing fruit crops, since commercial fruit growing is not traditional in Nepal. Most of the Nepalese farmers have small and fragmented holdings which are primarily occupied by cereals, and it is really difficult to persuade farmers to grow fruit trees instead of cereals.

Goverment Policies and Plans

Until 1990, both agriculture research and development were the responsibility of the Department of Agriculture under the Ministry of Agriculture. However, in 1991 the Nepal Agriculture Research Council (NARC) was formed as an autonomous organisation under a separate act. Now, NARC is mandated for all research activities and the Agriculture Department (DOA) for all extension and development activities.

There are many horticultural stations under DOA and NARC located in different agro-ecological regions of Nepal which are supposed to carry out both research and development activities. However, as far as research on deciduous fruits is concerned, very little work has been done. There are a number of reasons for that such as lack of qualified manpower and infrastructure, poor coordination between NARC and DOA, inadequate research funds, lack of long term planning and vision. Whatever research work has been carried out so far, it concentrated mostly on plant protection, cultivar collection and maintenance.

The Government has recognised horticulture as one of the important sub-sectors of agriculture in Nepal. Horticulture has been considered as a key to high growth rates and accorded priority in most of the past development plans. Despite the development potential and accorded priority, the pace of growth of horticulture has been very slow.

In view of the slow growth in the past and the ever increasing demand for horticultural products and to harness the immense potential of the country a "Twenty Year Horticulture Development Plan" was formulated in 1990 with support from the Asian Development Bank, named as the "Master Plan for Horticulture Development (MPHD)".

Despite the implementation of eight development plans in the past, no significant improvement in the living standards of the people has been achieved. Nonetheless, development efforts of the past decades have accelerated some basic infrastructure development. In view of this situation, a long term Agriculture Perspective Plan (APP) has been designed for Nepal that would lead past development towards a dynamic growth path. APP has taken agriculture as the engine of economic growth and that can trigger multiplier effects on the other sectors of the economy. Horticulture, sericulture and apiculture have been treated as high value commodities in the APP.

Based on the successful story of apple production in some areas and its advantages, the APP has identified apple as one of the high value priority commodities in high mountain areas. In the ongoing Ninth Five Year Plan, the Government has taken a policy of developing commercial orchards of apple in suitable pocket areas of western to far-western development regions. Eleven such districts have been recognised as most suitable for commercial apple cultivation.

Fruit Production in Bhutan

Bhutan has extremely diverse agro-climatic conditions due to major differences in altitude and rainfall as well as in slope characteristics. Roughly, the country could be divided into four physiographic zones - the southern foothills, the middle river valleys, the mountain slopes and the high Himalayas. Based on temperature and rainfall the country could be further sub-divided into six agro-climatic zones - wet subtropical, humid subtropical, dry subtropical, warm temperate, cool temperate and alpine zones.

Little is known about the history of deciduous fruit cultivation in Bhutan, as there is no documented evidence to draw information. The fact that traditionally fruits always occupied a prominent place on the altar of offering in all religious ceremonies lends support to the assumption that fruit cultivation must have been part of Bhutanese history, which could be traced as far back as the 7th century. Even now there are some existing orchards whose age could be more than 100 years. These orchards have all types of deciduous fruits like apricot, peach, pear, persimmon, walnut and chestnut and are found near almost all monasteries and district administration buildings in Bhutan.

Modern fruit cultivation however began in the reign of the third king Jigme Dorji Wangchuck, who is widely regarded in Bhutan as the Father of The Nation. It was his vision, which, as clearly stated in the foreword of his book on Planting an Orchard "Our country, the Kingdom of Bhutan, land of Buddhism, is a wonderful place. The variations in altitude and climatic conditions in different regions of Bhutan provide great scope for the growing of different fruits." In the absence of means of transport, however, fruits were grown in the past for domestic consumption only.

Prior to the inception of the Five-Year-Development Plan, fruit cultivation was limited to backyard cultivations only, meant solely for family consumption. With the beginning of the First Five-Year-Development Plan, fruit cultivation was identified as a potential source of cash income for the farmers and thus the first few commercial plantations were established. Apple was then the most important deciduous fruit crop introduced from India and grown on a commercial scale. Other deciduous fruits like apricot, peach, plum, pear and persimmon were still cultivated at subsistence level and were not of much economic significance.

Commercial cultivation of apple took some time to have significant economic impact owing to a relatively undeveloped road network and marketing systems. It was only in the late eighties that apple production had risen to the level of major commercial importance as a result of a better road network and the development of an export market to Bangladesh.

Deciduous Fruit Cutivation

Deciduous fruits grown in the backyards include Asian pear, peach, apricot, persimmon and walnut. Almost all of these have their indigenous species in the wild. The exotic and commercially cultivated deciduous fruits include apple, peach, plum, apricot, Bartlett pear, Nashi, and cherry. There is now a growing tendency to replace the old monastery orchards also with improved cultivars that are now available.

The main cultivars of apple are Red Delicious, Royal Delicious and Golden Delicious, which account for over 80% of the apple area. These cultivars were first introduced in the 1960s from the northern Indian States of Jammu & Kashmir and Himachal Pradesh. By the 1980s a host of new improved cultivars were brought in from Japan, which now constitute about 15% of the apple areas.

Apple is generally grown in temperate areas from about 1800m to 3000m. According to the preliminary survey carried out by the Integrated Horticulture Development Project in 1992, there were about 365,000 apple trees in the country, covering roughly about 1400 ha.

The present apple population is estimated to be around 565,000 based on the average annual sale of 50,000 plants from the Druk Seed Corporation in the last five years. Of this total, about 90% are found in the western part of the country covering the three Dzongkhags (districts) - Paro, Thimphu and Haa. The other 10% are found in the central district of Bumthang.

Unlike the beginning during the First Five Year Plan, 1962, apple growing has now become a major commercial venture. This is brought about vastly by the improved network of roads and marketing system from the late 1980s. Between 1960 and 1990 there has been a marked increase in the area under apple and other deciduous fruits.

The opening of the export market to Bangladesh in the late 80s has created awareness among the farmers of the three western Dzongkhags, who are the primary producers of apple. As a result, the cash income and the nutritional status of the farming communities of these three Dzongkhags has greatly improved. The three Dzongkhags together exported about 11,000 MT from estimated production of about 15,000 MT in 1997, for a value of Nu.169.6 million equivalent to US$ 4.24 million. The production is expected to increase to 23,000 MT in the next few years as the vast number of trees planted in the last decade come into bearing.

Planing Material

Traditional orchards found near the monasteries comprise of all seedling trees. The knowledge of clonal propagation of deciduous fruits came along only with the plants introduced from India. However, the country took some time to establish its own nursery industry and it was only in the last eight years or so that it could produce its own requirements of fruit plants.

Between 1960 and 1990, the annual fruit plant requirements were met by importing from nurseries in Himachal Pradesh and Uttar Pradesh, India. From 1991, the government nursery under the umbrella of the National Seed and Plant Production Programme was able to meet the national requirement of fruit plants. Initially this nursery used to produce and distribute its own plants but now it has limited its role to selling plants produced by its registered private growers.

Right from the start the technique of vegetative propagation was practiced. The rootstocks were propagated through stooling and hardwood cuttings while the scions were collected from mother trees. The most common form of grafting carried out was the whip and tongue grafting method and the most common budding method was the chip budding technique.

All species except walnut were propagated through grafting and budding. Until 1994, walnut was propagated through raising seedlings from seeds collected from bearing trees. From the winter of 1994 however, the first batch of walnut grafts were distributed to the farmers. These grafts were propagated through the Hot Callusing Device developed by H.B. Lagerstedt, USDA, Agricultural Research Service, Corvallis, Oregon, U.S.A.

Establishment of Orchards

The orchard establishment activities start with the fall season, when site clearing is done. Layout and pit digging is carried out in winter when the farmers have less field activities, as well as to allow sufficient time for the organic materials to decompose and the pit soil to settle. The recommended pit size is 3x3ft, and the most common layout followed is the triangular planting method. The standard spacing for the plants grafted on the semi-dwarfing clonal rootstocks is 3x3m, and for those on seedling rootstocks is 6x6m. Planting is done in late winter/early spring. Enough organic materials in the form of leaf mould and FYM mixed with recommended NPK fertiliser are added to the top soil set aside when digging each pit.

The newly planted trees are trained immediately after planting by heading back at 24 - 36 cm above ground. Subsequent annual pruning is carried out from the second year to build a good tree frame. The old orchards, however, did not receive good training and pruning as compared to the more recent plantings, owing to lack of technical backup support in the past. These orchards are now being gradually rejuvenated by top-working with improved varieties acceptable to the present export market needs.

Annual application of manures and fertilisers is done in winter with leaf mould, FYM and recomnended NPK at the time of basin preparation. Mulching is done immediately after, to preserve moisture and control weed growth in the growing season. Supplementary irrigation is carried out only in the research stations and a few demonstration orchards under a feasibility study programme.

With the increasing number of orchards the problem of pests and diseases has also been increasing. This is a serious factor that affects the productivity of the orchards.

Disease

The most serious disease which destroyed many orchards in the central district of Bumthang in mid 1980s is the Apple Scab (Venturia inaequalis). In the past, the control measures taken were severely hampered by the poor access to plant protection chemicals. The situation has improved with the establishment of the National Plant Protection Center (NPPC) in 1983 funded by the EEC. The

NPPC has now developed a Captan spray schedule to control scab.

The Powdery Mildew (Podosphaera leucotricha) is more damaging in young orchards as it affects the young shoots which if not treated in time leads to total destruction of the growing tips. Preventive measures are usually taken by spraying with Bavistin and Calyxin fungicides. Curative measures are taken by pruning-off the infected shoots and spraying with fungicides.

Collar Rot (Phytophthora cactorum) is associated with orchards with heavy and inadequately drained soil. It is also more pronounced in plants on MM106 rootstocks. Its incidence is controlled by adopting proper management practices of good drainage, proper planting method and good weed control.

Brown Rot (Sclerotinia fructigena) causes serious damage to fruits both on the trees as well as in storage. Infection usually starts with wounds caused by insects and mechanical injuries. Control involves spraying with Benzimidazole.

Peach Leaf Curl (Taphrina deformans) causes severe deformation of entire leaves on a peach plant. According to plant pathologists from the NPPC, the primary cause is aphid attacks followed by secondary infection by the fungus. Recommended control is spraying with copper oxychloride.

Apricot Brown Rot (Monilinia fruticola) causes severe losses to apricot production and is the single most damaging disease to the crop. It is controlled to a certain extent by benzimidazole.

Pests

The most serious pest which affects apple orchards is the Woolly Aphid (Eriosoma lanigerum). The control recommended for this pest is winter spraying with Lime Sulphur Liquid and Chlorpyriphos in the growing season.

Another serious pest which destroys the apple trees is the Stem and Twig Borer (Oberea posticata). This problem is pronounced in poorly managed and older orchards. Control is done by close observation and pruning the affected parts.

The adult of the Leaf and Fruit Feeding Beetles (Popilla sp, Hyperstylus sp, Microserica sp.) can seriously damage the foliage and fruit of most fruit trees. Complete defoliation may occur in young plants. The control spray should be timed to coincide with the feeding time of these beetles. Spraying with chlorpyriphos controls the infestation.

Trunk Borer (Anoplophora sp.) is the most damaging pest to the walnut trees, killing the plant outright if not detected early and controlled in time.

The use of plant protection chemicals is still very limited, owing to limited supply and strict regulations set by the National Plant Protection Center. Instead the center is carrying out studies at numerous sites to develop a sustainable Integrated Pest Management programme.

Harvesting and Yields of fruit

Harvesting of fruit starts with the early variety of cherry (Seneka) by the end of May. Next are the early varieties of peach in June, followed by the Japanese pear variety Hosui. The earliest variety of apple to be harvested is the Japanese variety Hana-iwai in July followed by Nebuta. The main apple harvest season starts from August and extends up to October. The last varieties of apple to be harvested are the Japanese varieties Mutsu and Fuji.

At present, apple dominates deciduous fruit production in Bhutan, followed by peach. The average yield of apple is relatively low owing to low plant density (250 trees/hectare), low level of management and less inputs. As a result, the average yield is reported at 45kg/ tree or 11 MT per hectare. Further, the quality of fruit is also variable owing to low level of management. The advantage associated with the Bhutanese apple, however, is the low usage of pesticides compared to the world's major apple growing countries.

Marketing

The production of apple on a commercial scale is a relatively recent development, primarily due to the development of roads and the export market. The export market is the single most important factor that resulted in the recent dramatic increase in the number of plantations; and it could absorb almost 100% of the apple produced, if not for the variable quality of the produce. Of the total annual production, about 30% of the produce does not meet the export requirements and hence is sold in the local market.

The Food Corporation of Bhutan (FCB) operates auction yards in three border towns close to the border with India, where the farmers take their horticultural produce including vegetables, and get good prices. The Agro-Food Processing Factory in Thimphu, established with the help of DANIDA, and located within the apple growing areas, constitutes a major local outlet not only for apples but also for other deciduous fruits. The regular Sunday market is the only form of wholesale market which is also a good retail outlet. At the moment, there are very few retail shops selling fruits primarily because of lack of good storage and packaging facilities.

At present, Bangladesh is the largest importer of Bhutanese fruits, importing about 70% of the total produce. The drawback with the Bangladesh market is that it demands apples of small to medium size only, which not only results in huge rejects of the common varieties like the Delicious group. At present, there are no markets for the large size fruit of improved Japanese varieties like Fuji, Sekaiichii and Mutsu. Therefore, there is a growing concern among the growers of the large sized apples, and demand on the government to explore means and ways to market those fruits to other markets in the region. Hopefully, the on-going UNDP assisted Integrated Horticulture Development Programme may help ease this situation in the near future.

Processing of Fruit Products

Processing of fruit products in Bhutan is constrained by the lack of small-scale processing units to suit low volume productions in areas far away from the road points, the shortage of labour and difficulties in transportation. The main fruit processing facilities in Bhutan at the moment are the Agro-Industries Ltd. in Thimphu, and the Bhutan Fruit Products Ltd. in Samchi. These two factories produce a wide range of fruit juices, jams, jelly, fruit cocktails, peach and apricot halves, cordials and similar products. The products are of high quality and are mainly marketed to major cities in India and Bangladesh.

Potential and Constraints

Bhutan has a huge potential to produce several deciduous fruit crops due to a wide range of agro-climatic conditions. Apart from growing apples and other deciduous fruits in the temperate zones, it could also grow peach, pear, apricot and grapes in its warm temperate zones, where there exists ample land resources to be exploited for this purpose.

If production is increased and quality improved, it could enjoy the benefits of close markets in India, Bangladesh and other countries in the region. Bhutan has the potential for the production of high quality produce in a clean environment; and the production seasons are different from most countries in the region. The other important factor is that the farmers are now aware of the benefits of growing fruit crops and are willing to cooperate in production development programmes. Horticultural produce has provided the best opportunities for the farmers to earn cash income, which is expected to provide further impetus to develop fruit crops like peach, pear, apricot, plums, cherries and grapes.

The major physical constraints to fruit production development are primarily due to the mountainous nature of the country. The narrow valleys with steep slopes restrict large-scale production of fruits by limiting mechanisation. The road network is still not well-developed, restricting transport of goods and produce from the farm to the market. In addition to the inadequate road network, horticultural production is constrained by limited infrastructure and facilities in terms of marketing, grading and packaging, storage and processing. As a result, there is limited access to the export markets outside India and Bangladesh.

Another limiting factor is at the institutional level, where fruit cultivation received only recent emphasis owing to late exploration of markets and also because the priority in the past was more on the production of grains to meet the national goal of food self-sufficiency. This has resulted in a shortage of trained personnel in the horticulture sector, less research activities and poor extension services.

In addition, information on horticultural production, markets and trade is limited and not reliable, leading to uncertainties as to the real commercial potential of different crops, and making effective policy and strategic planning difficult.

At the field level, yields are often low due to such problems as lack of knowledge among farmers in fruit production, poor varieties, low quality planting material, poor orchard management practices, and pest and disease problems.

In summary, the exploitation of the full potential of deciduous fruit cultivation in the country has not been possible due to the following constraints:

— Lack of clear-cut horticulture development policy and strategy in the past.

— Inadequate transportation network, including lack of feeder roads to potential production areas.

— A weak horticulture research and extension system.

— Limited farmer knowledge of fruit production systems.

— Insufficient market information and market facilities.

— Low yields and poor quality fruits due to low level of management practices.

— Poor post-harvest practices and lack of storage facilities.

— Lack of small-scale processing facilities.

Government Policies and Plans

As mentioned earlier, the horticulture development policy is based on the wisdom and the vision of the third King. This national commitment is the key to long-term success in horticultural development, and is particularly relevant in the attainment of the national development goal in the horticulture sector. Its aim is to increase production of horticultural crops both for domestic and export markets, leading to higher nutrition intake, higher farm incomes, and greater economic growth and export revenues.

The long term and short term objectives of horticultural development are:

Long Term Objective: "To optimise the contribution of the horticulture industry to the welfare of present and future generation of Bhutanese farmers and consumers with emphasis on improved cash income, food security, higher nutritional standards and export earnings".

Short Term Objective: "To improve the existing management practices and to increase production and quality of those horticultural commodities with a comparative advantage in a sustainable manner in all production environments through interdisciplinary research, extension and marketing".

To meet the objectives the following strategies are being adopted:

— Production based on market requirements.

— Production of high quality commodities.

— Production of high value produce with low perishability.

— Extension of supply periods through use of different varieties growing environments.

— Introduction and development of improved varieties.

— Intensification of production systems.

— Development of small-scale agro-industries.

— Development of kitchen gardens.

— Improvement of institutional capabilities.

To achieve the objectives and strategies, the following activities are being carried out:

— Applied and adaptive research.

— Strengthening extension services.

— Development of market information services and infrastructure.

— Human resources development.

— Improving the input supply system.

— Supporting private sector activities in the fruit industry.

Until the initiation of the first Integrated Horticultural Development Project (IHDP-I), funded by the UNDP and executed by FAO from 1990-1994, there was no systematic research in deciduous fruit cultivation. This project however, assisted in establishing a basic infrastructure and an action programme for horticulture research and development by studying production and market prospects, and strengthening the national capacity through training, germplasm introduction and technical assistance.

The main output of the project has been the preparation of a comprehensive long-term masterplan which has culminated into the on-going UNDP assisted Integrated Horticultural Development Programme (IHDP-II). This programme is geared towards an all-round development of horticulture covering all aspects such as production, research, extension, post-harvest activities and marketing. Its implementation commenced in July 1997. The programme is aimed at establishing a viable and vibrant horticultural industry in the long run.

The Integrated Horticulture Development Programme (IHDP-II) consists of six sub-programmes - a) Coordination sub-programme; b) Technology generation sub-programme; c) Extension sub-programme; d) Post-harvest sub-programme; e) Marketing sub-programme; and f) Aromatic and Medicinal Plants sub-programme.

— The Coordination sub-programme has the overall responsibility to ensure smooth functioning of the other sub-programmes.

— The Technology generation sub-programme looks after the research activities of horticulture in terms of introduction of germplasm, establishment of adaptive trials and generation of appropriate technologies.

— The Extension sub-programme works towards strengthening the horticulture extension activities.

— The Post-harvest sub-programme introduces post-harvest practices like grading, proper packaging and storage facilities.

— The Marketing sub-programme collects information on markets, explores new markets and develops markets for the horticultural products.

— The Aromatics and Medicinal sub-programme looks after the development of indigenous aromatic and medicinal plants and their uses.

5

Processing of Horticultural Crops

Processed products must be packaged and stored properly in order to achieve their potential shelf life of up to one year. Dried products must be packaged in airtight containers (glass or plastic bottles or sealed plastic bags). Canned and bottled products must be adequately heat processed using high quality containers that provide good seals. Dried and canned or bottled products are best stored in a cool, dark place.

Postharvest handling, transport and marketing of processed products can be much simpler and less costly than for fresh products, since refrigeration is unnecessary. Dried products take up much less space than their fresh equivalents, further reducing transport and storage costs.

Postharvest Processing Equipment

A catalog of postharvest processing equipment is available from Intermediate Technology Publications. Included are driers, storage containers, cleaners, hand mills, power mills, shellers, decorticators (seed removers), oil processing equipment, fruit presses, and root crop cutters/graters. Some examples are shown below.

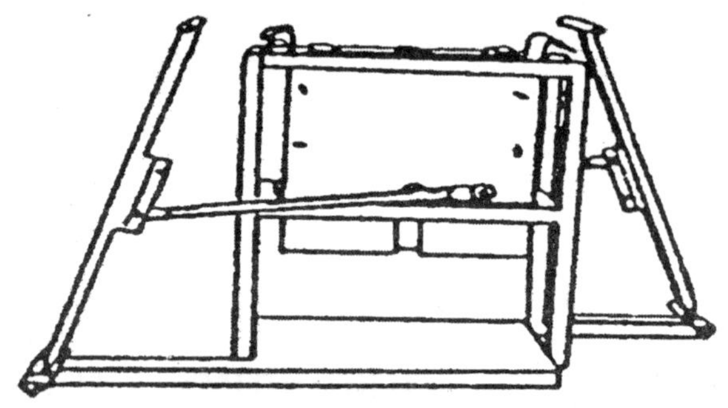

Two-man cassava grater

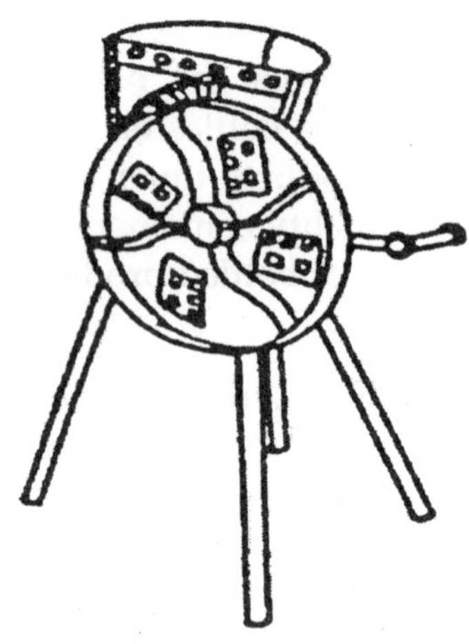

Four-bladed root chopper

Hand-operated fruit press

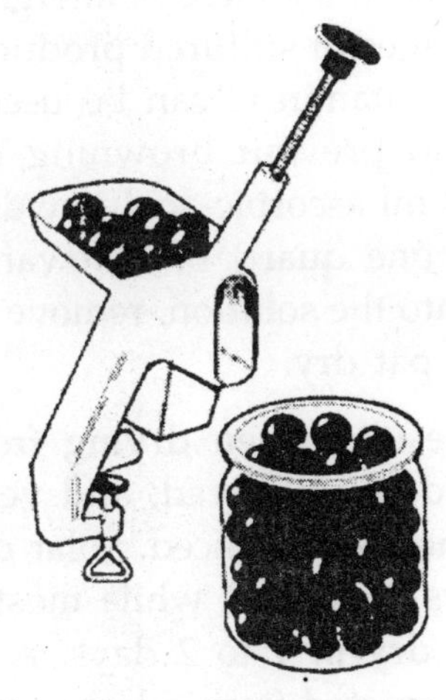

Cherry pitter

Preparation

Some produce requires blanching before freezing or drying. Blanching by boiling water bath or in steam ends certain enzymatic reactions in the product and helps retain color and flavor after processing. Always rinse blanched produce under very cold water or dip blanched produce into an ice water bath to stop the cooking process and quickly bring the temperature down.

Blanching times for selected commodities {use one gallon of water per pound (8 liters per kg) of produce}. Add one minute for each 2000 ft altitude if you live at elevations over 4000 ft. Fruits such as apples, pears, peaches and apricots are sometimes treated with sulfur being dried. Sulfuring {burn one tablespoon of sulfur powder per pound (12g per kg) of fruit} or sulphiting (dip fruit in a 1% potassium metabisulfite solution for one minute) helps prevent darkening, loss of flavor and loss of vitamin C.

Sulfur has been a source of allergic reactions in some people, so packages of sulfured product should always be clearly labeled. Vitamin C can be used as an alternative pre-treatment to prevent browning during the drying process. Use 30 ml ascorbic acid powder in one liter (or 2 tablespoons in one quart) of lukewarm water. Slice the fruits directly into the solution, remove with slotted spoon, drain well and pat dry.

For best results when drying fresh produce, fruits should be sliced or quartered, and vegetables should be thinly sliced, chopped or diced. Solar drying of fruits will take 2 to 3 days or longer, while most chopped or diced vegetables will dry in 1 to 2 days. A low cost sulfuring box can be constructed from a large cardboard box that is slashed in several places to allow adequate ventilation. Trays for drying be stacked using bricks and wooden spools as spacers. The trays must be made completely of wood, since sulfur fumes will corrode metal The entire assembly must be located out of doors, preferably on bare soil. Use one tablespoon of sulfur powder per pound (35 mls per kg) of fruit. Place the sulfur in a container well away from the side of the box since it will become quite hot. Seal the bottom edges of the box with soil.

Solar Drying Method

Horticultural produce can be dried using direct or indirect solar radiation. The simplest method for solar drying is to lay produce directly upon a flat black surface and allow the sun and wind to dry the crop. Nuts can be dried effectively in this way.

Simple direct driers can be made from trays of screening material propped upon wooden or concrete blocks to allow air to circulate under the produce. A layer of cheesecloth can be draped loosely over the produce, protecting it from insects and birds while drying. Check the produce each day, and move it under cover if rain threatens.

A simple method for solar drying is to construct a raised platform from wood and cover the frame with loosely woven mats. In the illustration below, sliced fresh tomatoes are being dried in direct sunlight on straw mats. Air can pass over and below the produce, speeding drying and reducing losses due to overheating. Aluminum foil can be utilised to reflect the sun onto the drying trays. The example shown below uses a plastic sheet to trap some of the heat and speed drying time.

All the trays, screens or mats used for drying produce must be kept clean. Trays made from stainless steel, plastic or nylon are much easier to keep clean than are wooden trays. Some fruits juices will adhere to the surfaces with each handling. This leads to an accumulation of dirt and buildup of mold that can contaminate and affect the appearance of the dried product. Use a strong detergent and a still brush to scrub the trays, screens or mats clean. Allow them to dry in the sun before using them to dry product.

A good material for settling dust on roads, paths or earthen floors is calcium chloride. When spread on the earth it absorbs moisture from the air and keeps the soil moist. Rake the calcium chloride into the surface at a rate of ½ pound per square yard. To reduce mold growth on trays, screens and mats during the off season, wash and dry them thoroughly, then store them in a well ventilated area.

In order to improve the efficiency of drying, some sort of structure must be used to capture solar radiation.

Direct Solar drier:

Indirect driers are constructed so the sun shines upon a solar collector (a shallow box, the insides painted black, topped with a pane of glass) heating air which then moves upward through a stack of four to six trays loaded with produce.

Forced-air Dehydrators

Nut crops can be dried in bulk using a dehydrator that combines a steady stream of air with an external source of heat. The plenum chamber below the produce is covered with a floor of perforated sheet metal or wooden slats. A fan located between the furnace and the plenum chamber moves the hot air through the drying produce.

Oil-burning Dehydrators

The batch-dryer illustrated below is constructed of wood, has an axial type fan and burns kerosene or diesel oil. A wide variety of dryers are available from manufacturers around the world.

Two types of dehydrators are commonly used for drying small volumes of nut crops. A wagon with a perforated floor can be transported from the field and connected to a portable burner batch drying. A stationary "pot-hole" dehydrator is designed to move heated air along a plenum under a fixed platform Individual bins of nuts are placed upon the platform and are dried as heat rises up through the perforated floor.

Wagon Dehydrator:

Pot-Hole Dehydrator:

Simple Electric Dehydrators

A simple electric dehydrator can be constructed using plywood, sheet metal a small fan, five household light bulbs with porcelain mounting fixtures and some screening material. The drier shown below is 32 inches long by 21 inches wide and 30 inches high, and contains racks for five trays. The fan and the sheet metal lining the bottom compartment help conduct heat upward through the box.

Oven Drying: Fruits and vegetables can be dried in a home oven if the oven can be run at a low temperature. Place the prepared produce on baking or metal screen trays, set the oven temperature at 140 °F and leave the door ajar (2 to 4 inches). Drying time can be reduced if ventilation is increased by using a small fan placed outside the oven.

Drying Flowers: Cut flowers can be air dried by hanging upside down or while supported by chicken wire. Certain flowers will look more natural if left standing in a vase while they dry. Anthurium dries best if left to dry very slowly. Cut the stems at a sharp angle, and place the flowers into a vase containing two inches of water. In all cases, flowers should be left to air dry in a dark, well area.

Flowers that dry best if left standing: strawflower, delphinium, larkspur, okra pods. Flowers that dry best while hanging upside down: chysanthemum, amaranthus, African daisy, statice, marigold African daisy supported on a screen of chicken wire:

Cut flowers can be dried quickly and easily in sand or silica gel. Sand used for drying flowers should be clean, smooth and the finer the better. Starting with one inch of sand in a container, place the flower to be dried on the sand and gently cover the entire flower with more sand. The container should be left uncovered and flowers should be dried in about three weeks. Flowers that dry well in sand are shasta daisy, lily-of-the-valley, cosmos, dahlia, sweet william carnation, stock, freesia and narcissus.

Drying Flowers in Sand: Silica gel is relatively expensive but reusable if heated to dry out the gel between uses. To use, cover the flower as with sand, then tightly seal the container. Check for drying in two to three days. Silica gel is especially useful for drying fragile plants and flowers with delicate colors. Flowers that dry best in silica gel are allium, anemone, cornflower, roses, tulip and zinnia.

Types of Canners

Two types of canners are commonly used to process horticultural crops. The first is a water bath canner, which is a large pot with a loose cover and a rack to hold jars off the bottom The pot should be deep enough to cover the canning jars by one to two inches and still have another inch of space to allow brisk boiling. The diameter of the pot should be no more than four inches wider than the diameter of the stove's burner to ensure even heating.

Acidic foods such as fruits, tomatoes, pickles and relishes, and high sugar foods such as jams, jellies, syrups and marmalades can be safely processed using a boiling water bath. A pressure canner is recommended for processing low acid foods such as vegetables. A pressure canner is a specially made heavy pot with a locking lid, an inner rack and a steam vent in the kid. The vent can be adjusted using a weight, value or screw, depending on the type of canner. A pressure gauge registers the air pressure inside the canner. A dial gauge gives a reading of the actual pressure, while weighted gauges will rock gently when the canner is at the proper pressure. Ten pounds of pressure at 115 °C (240 °F) is recommended for canning vegetables.

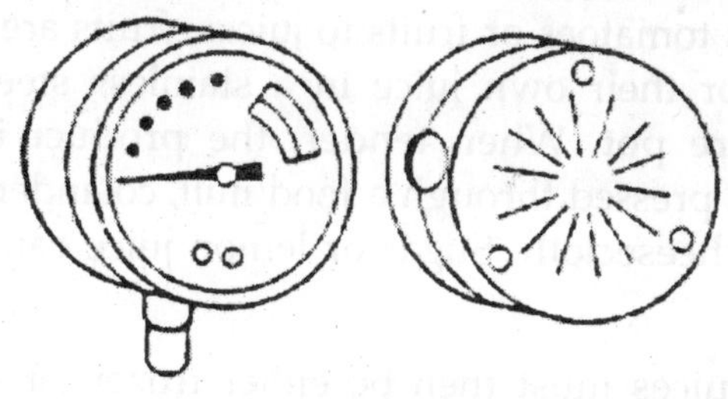

Dial Gauge

15 pounds pressure

5 pounds pressure

10 pounds pressure

Weighted Gauges

There are three types of glass canning jars used for processing horticultural crops. The ball type jar and the zinc capped jar both require rubber rings as seals. These can sometimes be difficult to obtain, but if locally available, make excellent containers. Currently the canning jar with a two-pieced lid is the most commonly used container.

No matter which jar is used, when filling containers, it is important to leave a small amount of headspace to allow for expansion of the food while processing. If a jar is filled too full, it may explode. If too much headspace is left, the food may spoil, since all the extra air may not be driven out during processing.

Juicing Fruits and Vegetables

To process tomatoes or fruits to juices, fruits are simmered in water or their own juice in a stainless steel, glass or enamelware pot. When tender, the product is cut into pieces and pressed through a food mill, colander or several layers of cheesecloth. Sugar or lemon juice can be added, to taste.

The juices must then be either frozen or canned for storage. Juices can be frozen in jars or freezer containers (leave 1/2 inch headspace). Most fruit juices can be canned in a boiling water bath for 20 minutes, but apple and grape juices can be processed in hot water (82 °C or 180 °F) for 30 minutes.

Vegetables should be chopped or shredded, then simmered for 45 to 50 minutes until mushy. The juice can then be pressed or strained from the vegetable pulp, and frozen or canned. Canning vegetable juices requires processing at ten pounds of pressure in a pressure canner. Pints should be processed for 55 minutes, and quart jars for 85 minutes.

Other Methods

Freezing : Most vegetables should be blanched before freezing to prevent loss of flavor and color during storage. Freezing temperatures are best set between -21 to -18 °C (0 to 5 °F). Packages for freezing should be moisture proof and vapor proof and contain as little air as possible to prevent oxidation during storage. Heavy plastic bags, heavy aluminum foil, glass freezer jars and waxed freezer cartons all make good containers.

Jellies, Jams and Preserves : Making jams, jellies and other high sugar preserves requires a balance of fruit, acid, pectin and sugar for best results. Under-ripe fruits contain more pectin than ripe fruits, and apple juice is a good source of natural pectin. If fruits are low in acid, lemon juice can be added to the mixture of fruit and sugar. Cane or beet sugar is better for making preserves than honey or corn syrup. To preserve fruits, cook on medium heat until the mixture "sheets" from a spoon. Avoid overcooking since this will lower the jelling capacity of the mixture. Pour into containers and seal with paraffin wax. The other preserves should be processed in a boiling water bath for five minutes.

Fermentation : When lactic acid bacteria in foods convert carbohydrates to lactic acid, food is preserved by the resulting low pH. Sauerkraut (cabbage) and wine (grapes) are two examples of thousands of fermented foods made around the world.

Acidification : Pickling is a simple processing method that can be used with many types of fruits and vegetables. Brine solution is poured over the product into glass canning jars (leave 1/2 inch headspace). Brined pickles are sealed and left at ambient temperature for three or more weeks, while fresh pack pickles are processed in a boiling water bath for 10 minutes.

Food Safety

Concerns about food safety when handling fresh fruits and vegetables have increased over the past decade. Recent outbreaks of food-bourne disease have been associated with berries, tomatoes, leafy greens and cut fruits. Wholesale buyers and consumers are increasingly interested in the use of handling practices that will ensure food safety. It is the responsibility of growers and postharvest handlers to document their practices to protect fresh produce from contamination.

Retailers such as large supermarket chains are demanding compliance with food safety practices from their suppliers. And soon it may be impossible to export produce to Europe or the U.S. without documenting its safe handling from the farm to the market. Some general food safety practices are being promoted by universities, governmental agencies and private sector organisations around the world. For growers who want to export their produce to the European Community, you should be aware that new standards are being developed by the retail produce industry to guide handling practices for growers and shippers (known as EUREP-GAP).

Key concepts are the implementation of GAP (Good Agricultural Practices) on the farm, in the packing house and during transport of all fresh produce, and HACCP (Hazard Analysis Critical Control Points) to document the safe handling of farm chemicals, pesticides, packaging materials, etc., especially for processed or fresh-cut produce. There is a EUREP-GAP deadline for food safety looming for fresh products intended for market in Europe. Export growers have until 2003 to meet specifications for hygiene and quality. Most small-scale packinghouses will require new water systems and improved sanitation practices. Current information can be found on the internet.

Guidelines for U.S. growers have been published by various sources. A copy of the publication: "Food safety begins on the farm: A grower's guide" can be obtained at no cost from Cornell University 's Good Agricultural Practices Programme. The typical causes and sources of food safety problems during production and postharvest handling fall into the following three major categories.

Physical Hazards: Examples of physical hazards which may become imbedded in produce during production handling or storage are such things as:

— fasteners (staples, nails, screws, bolts)

— pieces of glass

— wood splinters

Chemical Hazards: Examples of chemical hazards which may contaminate produce during production handling or storage are such things as:

— pesticides, fungicides, herbicides, rodentacides

— machine lubricants from forklifts or packing line equipment

— heavy metals (Lead, Mercury, Arsenic)

— industrial toxins

— compounds used to clean and sanitise equipment

Human Pathogens: There are four main types of human pathogens associated with fresh produce:

— soil associated pathogenic bacteria (Clostridium botulinum, Listeria monocytogenes)

— feces associated pathogenic bacteria (Salmonella spp., Shigella spp., E. coli O157:H7 and others)

— pathogenic parasites (Cryptosporidium, Cyclospora)

— pathogenic viruses (Hepatitis, Enterovirus).

Many of these pathogens are spread via a human (or domestic animal) to food to human transmission route. Handling of fruits and vegetables by infected field-workers or consumers, cross contamination, use of contaminated irrigation water, use of inadequately composted manure or contact with contaminated soil are just a few of the ways that transmission of human pathogens to food can occur.

While produce quality can be judged by outward appearance on such criteria as color, turgidity and aroma; food safety can not. Casual inspection of produce cannot determine if it is in fact safe and wholesome to consume. Management of growing and postharvest handling conditions are paramount in preventing the contamination of fresh produce by physical hazards, harmful chemicals and human pathogens.

Food Safety on the Farm

Practices related to these four simple principles can reduce the risk that produce may become contaminated on the farm.

Clean soil

— Avoid the improper use of manure.

— Compost manure completely to kill pathogens, and incorporate it into soil at least two weeks prior to planting.

— Keep domestic and wild animals out of fields to reduce the risk of fecal contamination.

— Provide portable toilet facilities* near the field.

— Prevent runoff or drift from animal operations from entering produce fields.

— Do NOT harvest produce within 120 days of a manure application.

Clean water

— Test surface water that is used for irrigation for fecal pathogens on a regular basis, especially if water passes close to a sewage treatment or livestock area.

— Keep livestock away from the active recharge area for well-water that will be used for irrigation.

— Keep chemicals away from the active recharge area for well-water that will be used for irrigation.

— Filter or use settling ponds to improve water quality.

— Where feasible, use drip irrigation to reduce crop wetting and minimise risk.

— Use potable water for making up chemical pest management sprays.

Clean Surfaces

— Tools and field containers must be kept clean. Wash and sanitise these items before each use.

Clean hands

— Workers who harvest produce must wash their hands after using the toilet.

— Provide soap, clean water and single-use towels in the field and insist that all workers wash their hands before handling produce.

Pathogen Contamination During Harvesting

During harvesting operations field personnel may contaminate fresh fruits and vegetables by simply touching them with an unclean hand or knife blade. Portable field latrines as well as hand wash stations must be available and used by all harvest crew members. Monitoring and enforcement of field worker personnel hygiene practices such as washing hands after using the latrine are a must, to reduce the risk of human pathogen contamination. Workers who are ill with hepatitis A or who have symptoms of nausea, vomiting or diarrhea should not be assigned to harvest fresh produce.

Produce once harvested should not be placed upon bare soils before being placed in clean and sanitary field containers. Field harvesting tools and gloves should be clean, sanitary and not be placed directly in contact with soil. Field containers should be cleaned and sanitised on a regular basis as well as being free of contaminants such as mud, industrial lubricants, metal fasteners or splinters. Do not allow workers to stand in field bins during harvest to reduce pathogen spread by shoes.

Plastic field bins and totes are preferred to wooden containers since plastic surfaces are easier to clean and sanitise, which should be done after every use. If containers are not cleaned and sanitised after every use, they may become contaminated and then contaminate the next products which are placed in the container. Wooden containers or field totes are almost impossible to sanitise since they have a porous surface and wooden or metals fasteners such as nails from wooden containers may accidentally be introduced into produce. Cardboard field bins if reused should be visually inspected for cleanliness and lined with a polymeric plastic bag before reuse to prevent the risk of cross contamination.

Depending upon the commodity, produce may be field packaged in containers that will go all the way to the destination market or be temporarily placed in bulk bins, baskets or bags which will be transported to a packing shed. Employees, equipment, cold storage facilities, packaging materials and any water which will be contacting the harvested produce must be kept clean and sanitary to prevent contamination.

Minimising Pathogen Contamination During Postharvest Handling

Gloves, hairnets and clean smocks are commonly worn by packinghouse employees in export oriented packing sheds. The cleanliness and personnel hygiene of employees handling produce at all stages of production and handling must be managed to minimise the risk of contamination. Adequate bathroom facilities and handwash stations must be provided and used properly to prevent contamination of produce by packinghouse employees. Shoe or boot cleaning stations may also be in place to reduce the amount of field dirt and contamination which enters the packing shed from field operations.

Food contact surfaces on conveyor belts, dump tanks etc. should be cleaned and sanitised on a regular scheduled basis with food contact surface approved cleaning compounds. A 200 parts per million sodium hypochlorite (bleach) solution is an excellent example of a food contact surface sanitiser. Sanitisers should be used only after thorough cleaning with abrasion to remove organic materials such as dirt or plant materials. Use of steam to clean equipment should be avoided since steam may actually cake organic materials and form a biofilm, which renders equipment almost impossible to sanitise. Steam may also aerosolise bacteria into the air and actually spread contamination throughout the packing house facility.

All packaging materials should be made of food contact grade materials to assure that toxic compounds in the packaging materials do not leach out of the package and into the produce. Toxic chemical residues may be present in some packaging materials due to use of recycled base materials. Empty packages such as boxes and plastic bags should be stored in an enclosed storage area to protect them from insects, rodents, dust, dirt and other potential sources of contamination. These actions protect not only against the potential loss of valuable materials but protects the integrity and safety of these materials.

All water which comes in contact with produce for washing or hydrocooling must be safe to drink. Water should contain between 100 and 150 parts per million total chlorine and have a pH of between 6 and 7.5. Chlorine use prevents the potential for cross contamination of all produce in the washing or hydrocooling system, it will not sterilise the produce. Change the water in dump tanks and hydro-coolers regularly.

Use ice made from only potable water sources.

Produce is best shipped in temperature controlled refrigerated trucks. Pre-cool the vehicles prior to loading. Maintaining perishables below 5°C (41° F) even while being transported to destination markets will extend shelf-life and significantly reduce the growth rate of microbes including human pathogens. Temperatures used for tranporting chilling sensitive produce will not protect against the growth of most pathogens. Trucks used during transportation should be cleaned and sanitised on a regular basis. Trucks which have been used to transport live animals, animal products or toxic materials should never be used to transport produce.

High pressure wash, rinse and sanitise all crop containers, tools and packhouse surfaces prior to each day's harvest. Sanitisers should be used only after thorough cleaning with abrasion to remove organic materials such as dirt or plant materials. Most commercial sanitisers contain chlorine or quaternary ammonium compounds (QUATS, QAC, benalkonium chloride, N-alkyl dimethybenzyl ammonium chloride). Chlorine solutions prepared from chlorine gas, hypochlorites and cloramines are not compatible with quaternary ammonium compound sanitisers.

Selection of the sanitiser to use depends upon the surface to be cleaned, hardness of the water, application equipment available, effectiveness under ambient conditions and cost. All require extreme care when handled as either compressed gas, powders or concentrated liquids. Use the following table to assist in the selection of appropriate sanitisers for your operation.

The ability to identify the source of any fresh produce product is an essential part of good agricultural practices. This is especially important whenever several growers are contributing to a single load, or when produce is being exported by one grower from several different fields.

The following practices are advocated by the U.S. Food and Drug Administration.

— Develop procedures to track produce from the farm, to the packer, distributor and retailer.

— Documentation should indicate the source of the product (farm identification and the field where it was produced), the date of harvest, and who handled the produce.

— Growers, packers and shippers should partner with transporters, distributors and retailers to develop management tools to facilitate the traceback process.

6

Storage of Horticultural Crops

Commodities stored together should be capable of tolerating the same temperature, relative humidity and level of ethylene in the storage environment. High ethylene producers (such as ripe bananas, apples, cantaloupe) can stimulate physiological changes in ethylene sensitive commodities (such as lettuce, cucumbers, carrots, potatoes, sweet potatoes) leading to often undesirable color, flavor and texture changes.

Temperature management during storage can be aided by constructing square rather than rectangular buildings. Rectangular buildings have more wall area per square feet of storage space, so more heat is conducted across the walls, making them more expensive to cool. Temperature management can also be aided by shading buildings, painting storehouses white or silver to help reflect the sun's rays, or by using sprinkler systems on the roof of a building for evaporative cooling.

The United Nations' Food and Agriculture Organization (FAO) recommends the use of ferrocement for the construction of storage structures in tropical regions, with thick walls to provide insulation. Facilities located at higher altitudes can be effective, since air temperature decreases as altitude increases. Increased altitude therefore can make evaporative cooling, night cooling and radiant cooling more feasible. Underground storage for citrus crops is common in Southern China, while in Northwest China, apples are stored in caves. This system was widely used in the U.S. during the early pert of this century.

Certain commodities, such as onions and garlic, store better in lower relative humidity environments. Curing these crops and allowing the external layers of tissue to dry out prior to handling and storage helps to protect them from further water loss.

The air composition in the storage environment can be manipulated by increasing or decreasing the rate of ventilation (introduction of fresh air) or by using gas absorbers such as potassium permanganate or activated charcoal. Large-scale controlled or modified atmosphere storage requires complex technology and management skills, however, some simple methods are available for handling small volumes of produce.

Storage Temperatures

Recommended Temperature and Relative Humidity, and Approximate Transit and Storage Life for Fruits and Vegetable Crops (see Hardenburg et al, 1986 for more complete information on individual crops).

Compatibility Groups for Storage

Group 1: Fruits and vegetables, 0 to 2°C (32 to 36°F), 90-95% relative humidity. Many products in this group produce ethylene.

apples	grapes (without	parsnips
apricots	sulfur dioxide)	peaches
Asian pears	horseradish	pears
Barbados cherry	kohlrabi	persimmons
beets, topped	leeks	plums

berries (except cranberries)	longan	pomegranates
cashew apple	loquat	prunes
cherries	lychee	quinces
coconuts	mushrooms	radishes
figs (not with apples)	nectarines	rutabagas
	oranges	turnips

Group 2: Fruits and vegetables, 0 to 2°C (32 to 36°F), 95-100% relative humidity. Many products in this group are sensitive to ethylene.

Amaranth*	corn, sweet*	parsley*
anise	daikon*	parsnips*
artichokes*	endive*	peas*
asparagus	escarole*	pomegranate
bean sprouts	grapes (without sulfur dioxide)	raddichio
beets*		radishes*
Belgian endive	horseradish	rhubarb
berries (except cranberries)	Jerusalem artichoke	rutabagas*
kiwifruit	salsify	
bok choy	kohlrabi*	scorzonera
broccoli*	leafy greens	snow peas
brussels sprouts*	leeks' (not with figs or grapes)	spinach*
cabbage*		turnips*
carrots*	lettuce	waterchestnut
cauliflower	lo bok	watercress*
celeriac*	mushrooms	
celery*	onions, green* (not with figs, grapes, mushrooms, rhubarb, or corn)	
cherries		

* these products can be top-iced

Group 3: Fruits and vegetables, 0 to 2°C (32 to 36°F), 65-75% relative humidity. Moisture will damage these products.

Garlic onions, dry

Group 4: Fruits and vegetables, 4.5°C (40°F), 90-95% relative humidity.

cactus leaves	lemons*	tamarillo

cactus pears	lychees	tangelos*
caimito	kumquat	tangerines*
cantaloupes**	mandarin*	ugli fruit*
clementine	oranges (Calif. and Arizona)	yucca root
cranberries	pepino	

* citrus treated with biphenyl may give odors to other products.

** can be top-iced.

Group 5: Fruits and vegetables, 10°C (50°F), 85-90% relative humidity. Many of these products are sensitive to ethylene. These products also are sensitive to chilling injury.

beans	kiwano	pummelo
calamondin	malanga	squash, summer (sot shell)
chayote	okra	tamarind
cucumber	olive	taro root
eggplant	peppers	
haricot vert	potatoes, storage	

Group 6: Fruits and vegetables, 13 to 15°C (55 to 60°F), 85-90% relative humidity. Many of these products produce ethylene. These products also are sensitive to chilling injury.

atemoya	granadilla	papayas
avocados	grapefruit	passionfruit
babaco	guava	pineapple
bananas	jaboticaba	plantain
bitter melon	jackfruit	potatoes, new
black sapote	langsat	pumpkin
boniato	lemons*	rambutan
breadfruit	limes*	santol
canister	mamey	soursop
carambola	mangoes	sugar apple
cherimoya	mangosteen	squash, winter (hard shell)
coconuts	melons (except cantaloupes)	tomatillos
feijoa		tomatoes, ripe
ginger root		

*citrus treated with biphenyl may give odors to other products

Group 7: Fruits and vegetables, 18 to 21°C (65 to 70°F), 85-90% relative humidity.

jicama	sweetpotatoes*	watermelon*
pears	tomatoes,	white sapote
(for ripening)	mature green	yams*

*separate from pears and tomatoes due to ethylene sensivity.

Group 8: Flowers and florist greens, 0 to 2°C (32 to 36°F), 90-95% relative humidity.

allium	freesia	peony, tight
aster, China	gardenia	buds
bouvardia	hyacinth	ranunculus
carnation	iris, bulbous	rose
chrysanthemum	lily	squill
crocus	lily-of-the-valley	sweet pea
cymbidium orchid	narcissus	tulip
adiantum (maidenhair)	ground pine	rhododendren
cedar	flex (holly)	salal (lemon leaf)
dagger and wood	juniper	
ferns	mistletoe	vaccinium
galax	mountain-laurel	(huckleberry)
woodwardia fern		

Group 9: Flowers and florist greens, 4.5°C (40°F), 90-95% relative humidity.

acacia	delphinium	orchid,
alstromeria	feverfew	cymbidium
anemone	forget-me-not	ornithogalum
aster, China	foxglove	poppy
buddleia	gaillardia	phlox
calendula	gerbera	primrose
calla	gladiolus	protect
candytuft	gloriosa	ranunculus
clarkia	gypsophilla	snapdragon
columbine	heather	snowdrop
coreopsis	laceflower	statice
cornflower	lilac, forced	stephanotis

cosmos	lupine	stevia
dahlia	marigolds	stock
daisies	mignonette	strawflower
violet	zinnia	
adiantum (maidenhair)	eucalyptus	myrtus (myrtle)
asparagus (plumosa, sprenger)	hedera	philodendren
	flex (holly)	pittosporum
buxus (boxwood)	leatherleaf (baker fern)	pothos
camellia		scotch-broomern
croton	leucothoe, drooping	smilax, southern
dracaena	magnolia	woodwardia fern

Group 10: Flowers and florist greens, 7 to 10°C (45 to 50°F), 90-95% relative humidity.

anemone	eucharis	orchid, cattleya
bird-of-paradise	gloriosa	sweet william
camellia	godetia	
chamaedora	cordyline (ti)	palm
podocarpus		

Group 11: Flowers and florist greens, 13 to 15°C (55 to 60°F), 90-95% relative humidity.

anthurium	heliconia	poinsetta
ginger	orchid, vanda	
diffenbachia	stag horn fern	

Cleaning Storage Practices

Inspecting stored produce and cleaning storage structures on a regular basis will help reduce losses by minimizing the buildup of pests and discouraging the spread of diseases.

Storage facilities should be protected from rodents by keeping the immediate area clean, free from trash and weeds. Rat guards can be made from simple materials such as old tin cans or pieces of sheet metal fashioned to fit the extended legs of storage structures. If desired, more elaborate technologies can be used. Concrete floors will help prevent rodent entry, as will screens on windows, vents and drains.

Remove trash and weeds

Rat guards

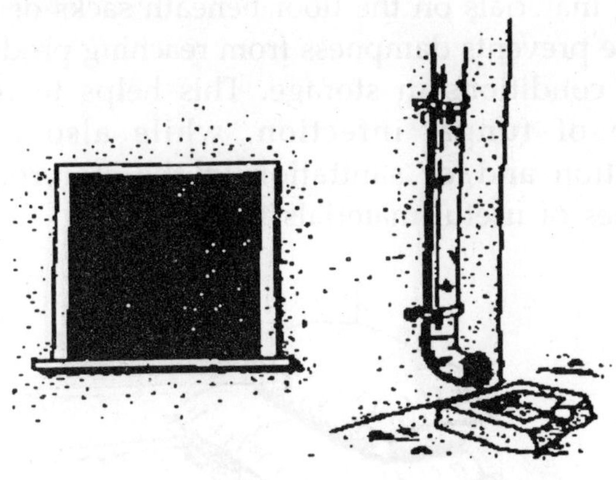

Screens

Cement floors

When inspecting stored produce, any spoiled or infected produce should be removed and destroyed. In some cases, produce may still be fit for consumption if used immediately perhaps as animal feed. Reusable containers and sacks should be disinfected in chlorinated or boiling water before reuse.

Placing materials on the floor beneath sacks or cartons of produce prevents dampness from reaching produce suited to dry conditions in storage. This helps to reduce the chance of fungal infection, while also improving ventilation and/or sanitation in the storeroom. Some examples of useful materials follow:

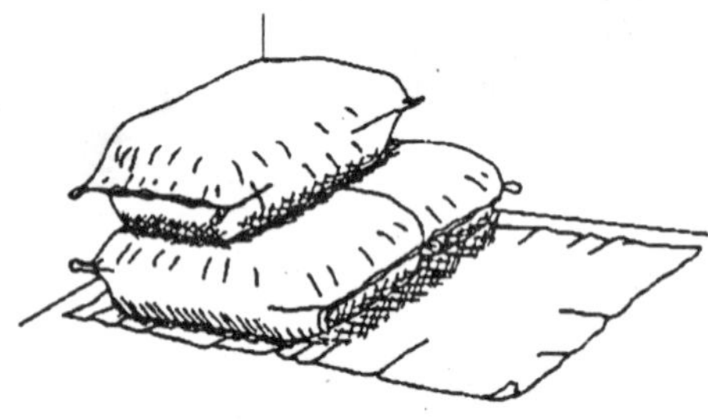

Waterproof sheets

Poles

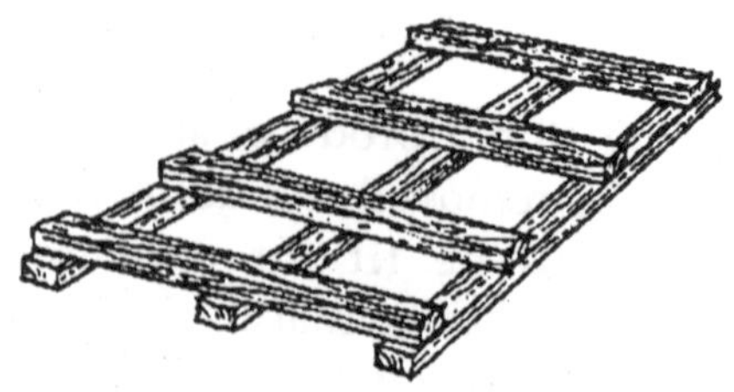

Wooden pallets

Structure of Storage

A yam barn is a traditional structure used in West Africa to store yams after curing. Fast-growing, live trees are used to create a rectangular structure, and form the framework of the barn as well as provide shade. Storage facilities require adequate ventilation in order to help extend shelf life and maintain produce quality. The following are three types of fans found in common use.

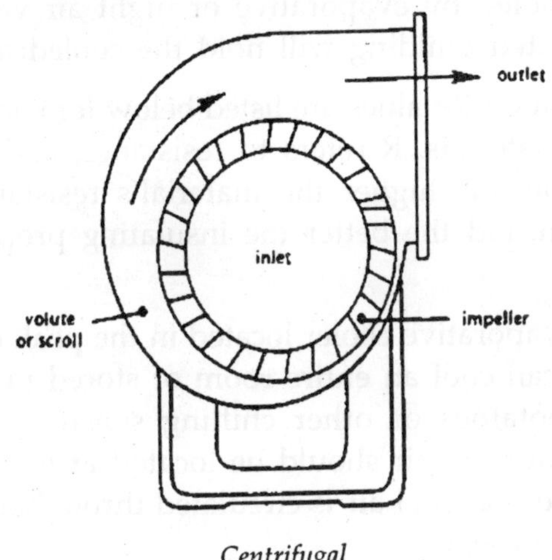

Centrifugal

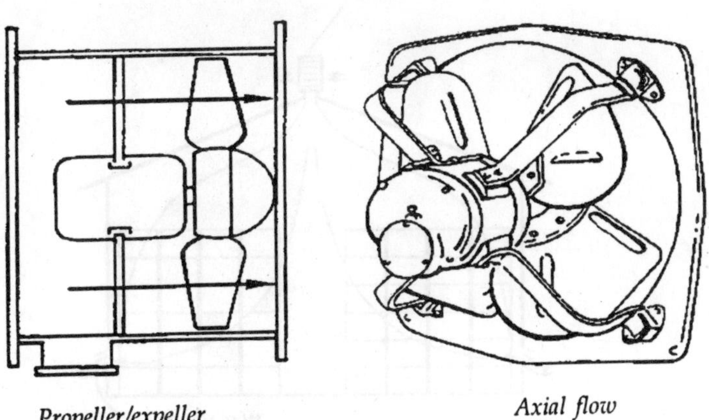

Propeller/expeller

Axial flow

Ventilation in storage structures is improved if air inlets are located at the bottom of the store, while air outlets are at the top. A simple, light-proof exhaust vent is a pressure-relief flap.

Any type of building or facility used for storage of horticultural crops should be insulated for maximum effectiveness. A well insulated refrigerated building will require less electricity to keep produce cool. If the structure is to be cooled by evaporative or night air ventilation, a well insulated building will hold the cooled air longer.

Insulation R-values are listed below for some common building materials. R refers to resistance, and the higher the R-value, the higher the material's resistance to heat conduction and the better the insulating property of the material.

An evaporative cooler located in the peak of a storage structure can cool an entire room of stored produce such as sweetpotatoes or other chilling sensitive crops. The vents for outside air should be located at the base of the building so that cool air is circulated throughout the room before it can exit.

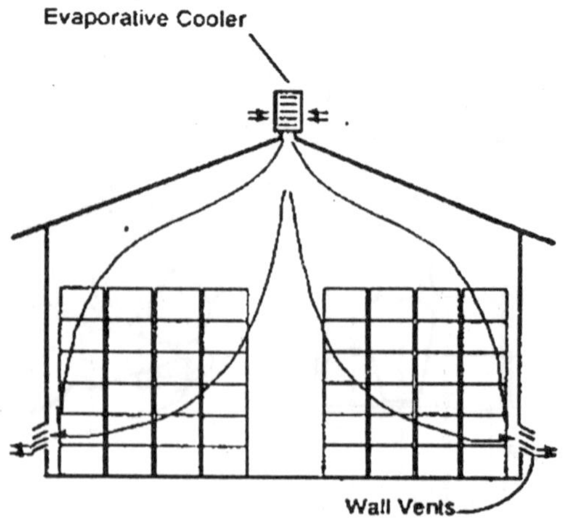

Illustrated below is a cross-sectional view of a storehouse for fruits. This system was officially approved as the standard model for farm-level storehouses by the Ministry of Construction, Korea, in 1983. Note that air inlets are at the base of the building, and the floor is perforated, allowing free movement of air. The entire building is set below ground level taking advantage of the cooling properties of soil.

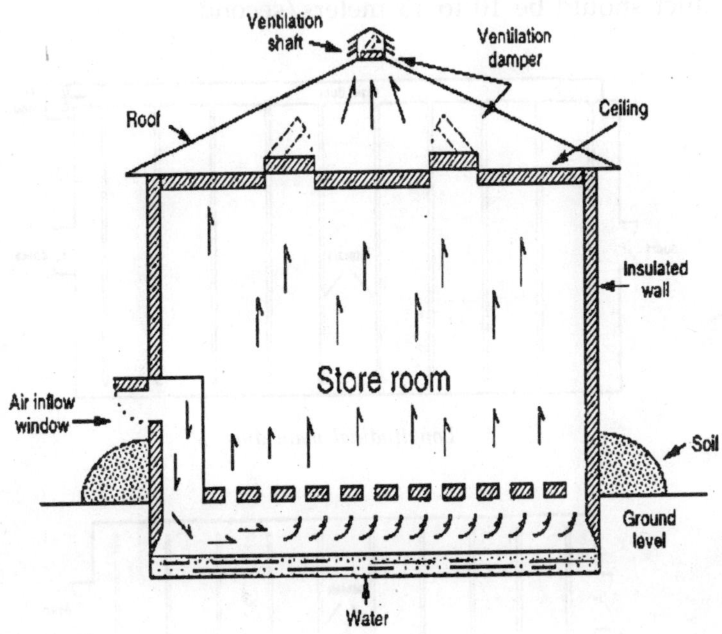

Commercially constructed cold rooms can be quite expensive, but fortunately the small-scale operator has many choices. Cold rooms can be self-constructed, purchased as prefabricated units (new or used), or made from refrigerated transportation equipment such as railway cars, highway vans or marine containers.

For storage facilities that are refrigerated, using outside air for ventilation is wasteful of energy. For these systems, a simple recirculation system can be designed by adding a fan below floor level and providing a free space at one end of the storeroom for cool air to return to the inlet vents. The proper arrangement of floor vents for air circulation will improve ventilation in the storage house. Lateral ducts should be 2 meters apart when measured from center to center. Air How velocities from the main duct should be 10 to 13 meters/second.

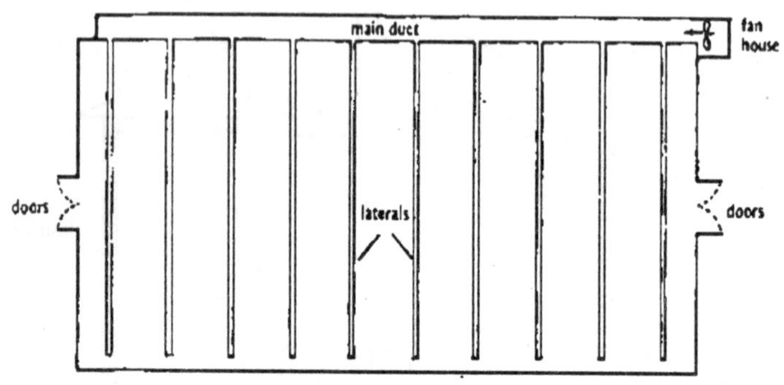

Longitudinal main duct

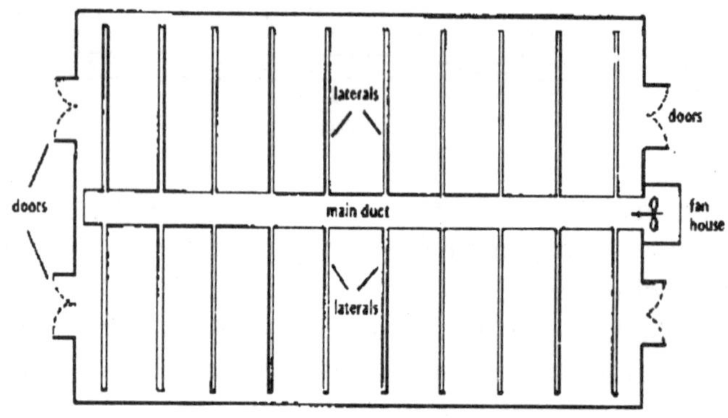

Central main duct

Lateral ducts can be constructed of a variety of materials. Portable vents can be made from wooden slats, in a triangular, square or rectangular design. A round tube of plastic or clay can be used if holes can be drilled without causing structural damage, or permanent ducts can be constructed below ground, using concrete blocks.

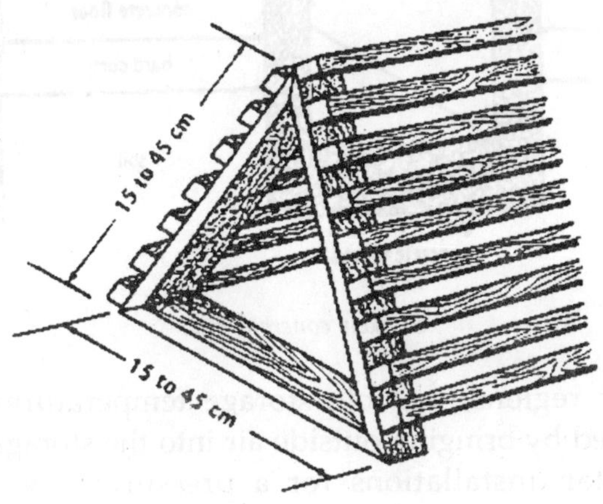

Triangular wooden duct:

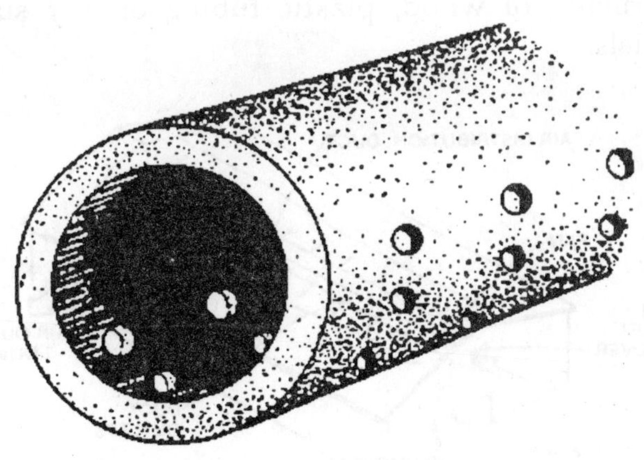

Clay pipe duct

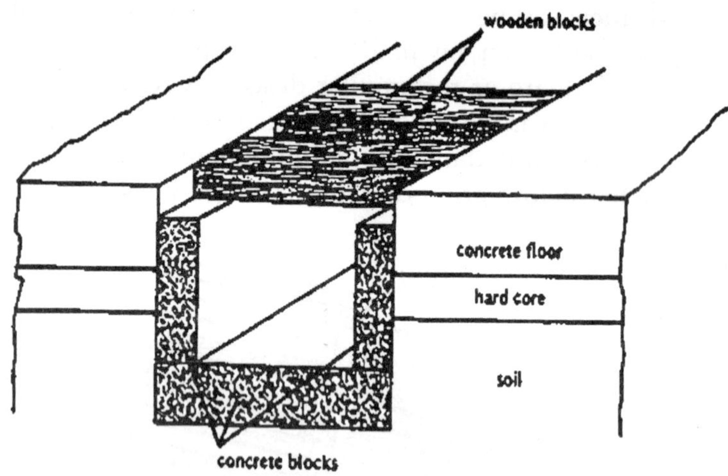

Sunken concrete duct

In cooler regions, suitable storage temperatures can be maintained by bringing outside air into the storage facility. Typical fan installations for a pressurized ventilation system are illustrated below. Overhead distribution of air simplifies the storehouse design. A indoor recirculation inlet can be added if refrigeration is in use. Ducts can be constructed of wood, plastic tubing or any suitable materials.

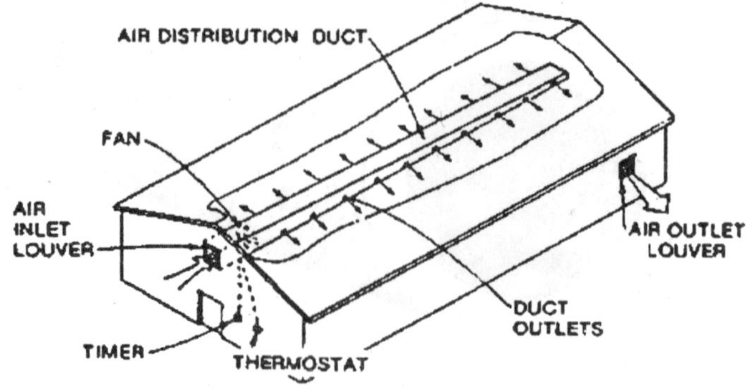

Overhead ventilation distribution system

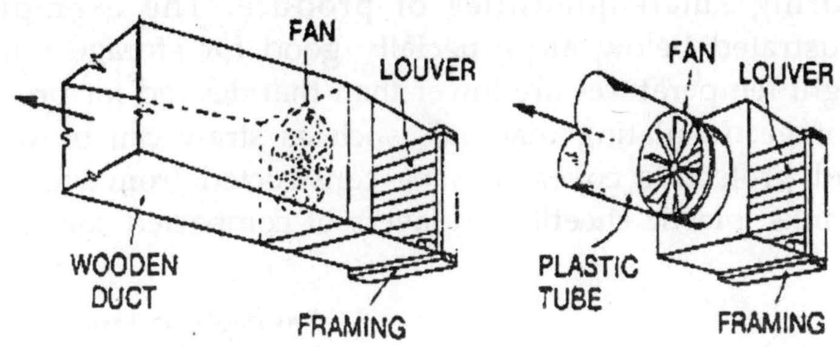

Types of ducts for air inlet fans

Storage structures can be cooled by ventilating at night when outside air is cool. For best results, air vents should be located at the base of the storage structure. An exhaust fan located at the top of the structure pulls the cool air through the storeroom. Vents should be closed at sunrise, and remain closed during the heat of the day.

Overhanging roof extensions on storage structures are very helpful in shading the walls and ventilation openings from the sun's rays, and in providing protection from rain. An overhang of at least 1 meter (3 feet) is recommended.

Where electricity is not available, wind-powered turbines can help keep storerooms cool by pulling air up through the building. Vents at the floor level are especially useful for cooling via night air ventilation.

The turbine can be constructed of sheet metal that is twisted to catch the wind, and attached to a central pole that acts as the axis of rotation. Warm air in the storage room rises, causing the turbine to rotate, expelling the air and initiating an upward flow of warm air. The turbine should be placed on the peak of the roof of a storage structure.

Protected surface storage is a simple method for storing small quantities of produce. The examples illustrated below are especially good for storage when night temperatures are lower than that desired for proper storage. Insulating materials such as straw can be used and protective covers can be constructed from wooden planks, plastic sheeting or layers of compacted soil.

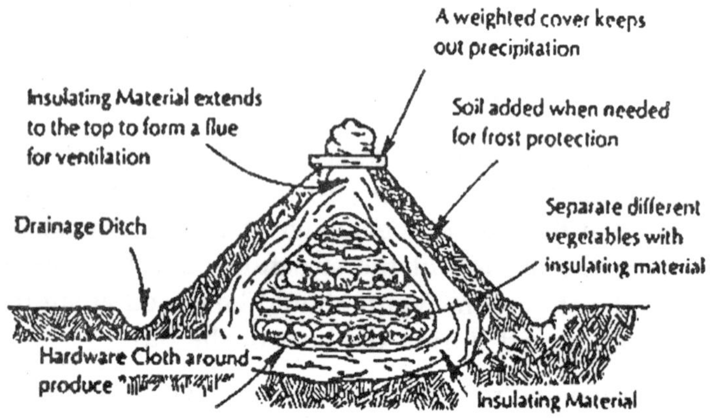

Cone-shaped pit storage

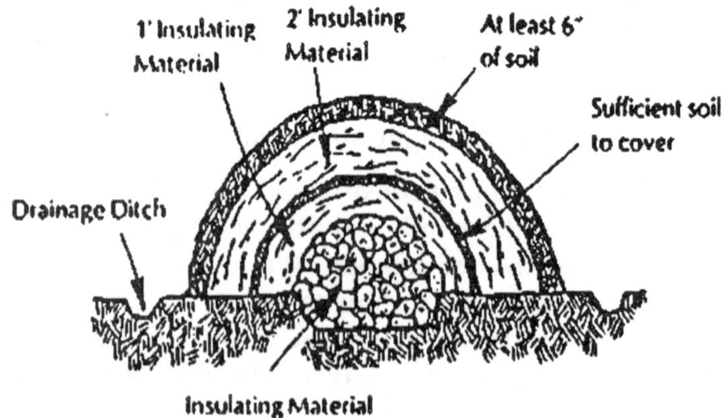

Mound storage

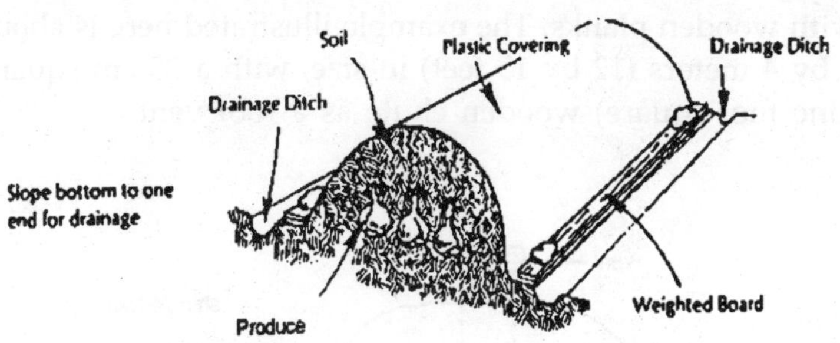

Trench storage

One of the simplest methods for storing small quantities of produce is to use any available container, and create a cool environment for storage by burying the container using insulating materials and soil. The example provided here employs a wooden barrel and straw for insulation.

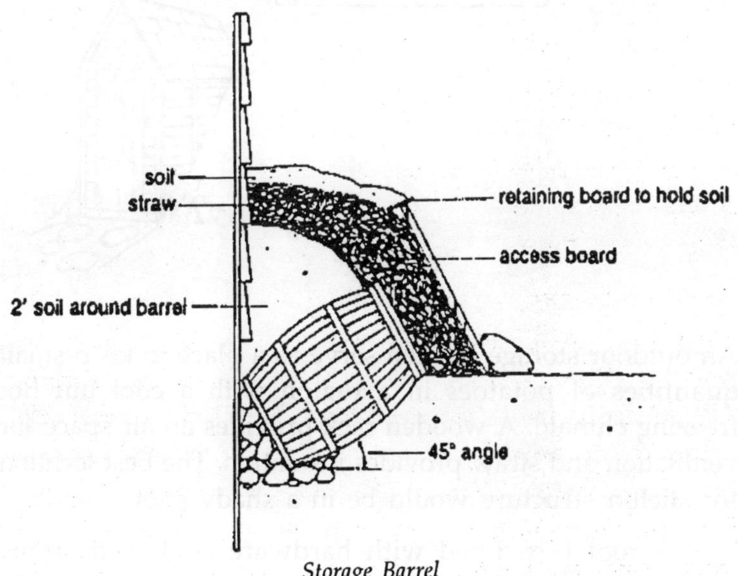

Storage Barrel

A root cellar can be constructed by digging out a pit to a depth of about 2 meters (7 to 8 feet) and framing the sides with wooden planks. The example illustrated here is about 3 by 4 meters (12 by 15 feet) in size, with a 35 cm square (one foot square) wooden chute as a roof vent

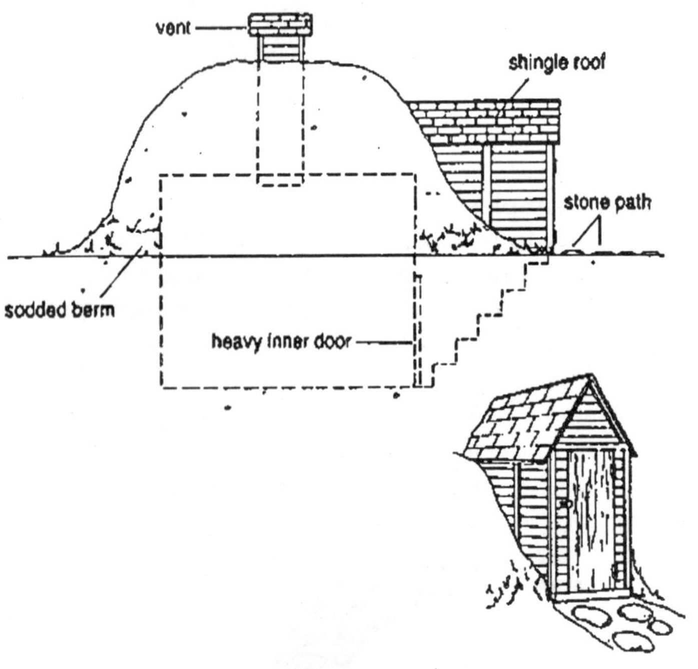

An outdoor storage bin can serve as a place to keep small quantities of potatoes in a region with a cool but not freezing climate. A wooden rack provides an air space for ventilation and straw provides insulation. The best location for such a structure would be in a shady spot.

A root box, lined with hardware cloth and straw, buried to the top edge in soil will keep potatoes cool wile providing protection from freezing. The wooden lid can be lifted for easy access to produce, and straw bales on top provide more insulation.

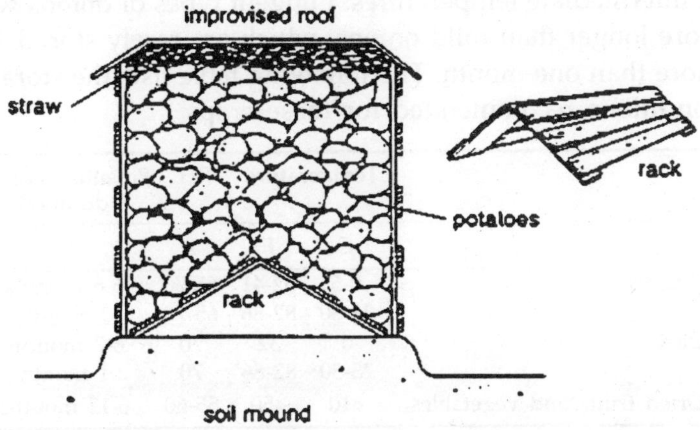

Storage Bin

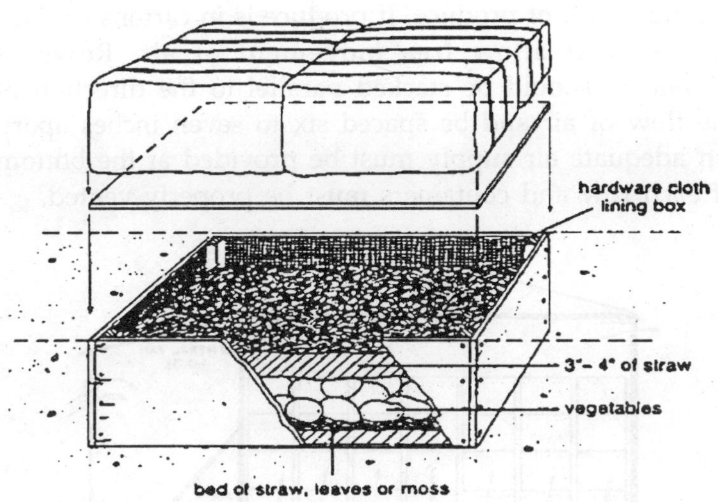

Root Box

Dried and Bulb Crops

Onions, garlic and dried produce are best suited to low humidity in storage. Onions and garlic will sprout if stored at intermediate temperatures. Pungent types of onions will store longer than mild onions, which are rarely stored for more than one month. The following table lists the storage conditions recommended for these crops.

	Temperature		RH	Potential storage duration
	C	F	%	
Onions	0-5	32-41	65-70	6-8 months
	28-30	82-86	65-70	1 month
Garlic	0	32	70	6-7 months
	28-30	82-86	70	1 month
Dried fruits and vegetables	<10	<50	55-60	6-12 months

For bulk storage of onions or garlic, ventilation systems should be designed to provide air into the store from the bottom of the room at a rate of 2 cubic feet per minute per cubic feet of produce. If produce is in cartons or bins, stacks must allow free movement of air. Rows of containers should be stacked parallel to the direction of the flow of air and be spaced six to seven inches apart. An adequate air supply must be provided at the bottom of each row and containers must be properly vented.

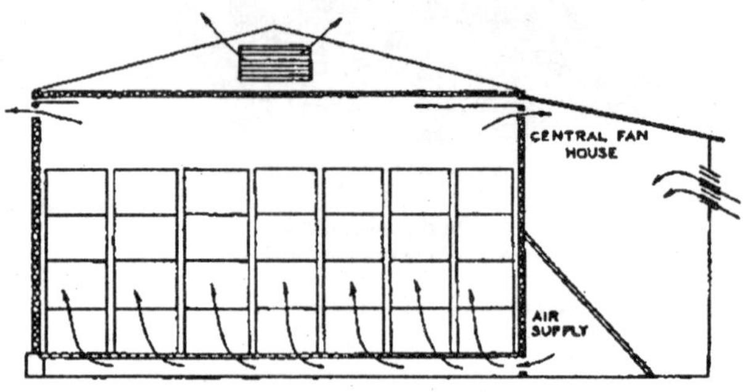

Storage in cartons or bins

Tropical Root and Tuber Crops

The recommended storage conditions for root and tuber crops are listed in the following table. Potatoes for processing are best kept at intermediate temperatures to limit the production of sugars which darken when heated during processing. Potatoes meant for consumption must also be stored in the dark, since the tubers will produce chlorophyll (turning green) and develop the toxic alkaloid solanine if kept in the light. Potatoes stored for use as "seed" are best stored in diffuse light. The chlorophyll and solanine that accumulate will aid to protect the seed potatoes from insect pests and decay organisms.

Tropical root and tuber crops must be stored at temperatures that will protect the crops from chilling, since chilling injury can cause internal browning, surface pitting and increased susceptibility to decay.

	Temperature C	F	RH(%)	Potential storage duration
Potatoes				
Fresh market	4-7	39-45	95-98	10 months
Processing	8-12	47-54	95-98	10 months
Seed potatoes	0-2	32-36	95-98	10 months
Cassava	5-8	41-47	80-90	2-4 weeks
	0-5	32-41	85-95	6 months
Sweetpotato	12-14	54-58	85-90	6 months
Yam	13-15	55-59	near 100	6 months
	27-30	80-86	60-70	3-5 weeks
Ginger	12-14	54-58	65-75	6 months
Jicama	12-15	54-59	65-75	3 months
Taro	13-15	55-59	85-90	4 months

Storing Potatoes

When storing potatoes, a field storage clamp is a low cost technology that can be designed using locally available materials for ventilation and insulation. The example illustrated here employs a wooden ventilator box and straw for insulation. The entire pile of potatoes and straw is covered with a layer of soil, which should not be highly compacted. In very cold regions, a second layer of straw and soil can be added. In hot regions, less soil is needed, but more ventilation can be added by constructing chimney type air outlets at the top of the clamp.

Simple storage houses for potatoes can be constructed for small quantities of produce. The examples provided here can store 1 to 2 metric tons, and are used on farms and in mountain villages. The first is made from unfinished wooden planks painted white to reduce heat accumulation from the sun and covered with a large thatched roof for protection from sun and rain. It has a large door on one side for loading and unloading.

The second storage house is constructed from lath and plaster and mud bricks in a cylindrical form. It has two doors, one on top for loading, the other at the bottom for easy removal of potatoes for sale or consumption. Whitewash helps reduce heat accumulation and a thatch roof protects the potatoes from rain and sun.

For large quantities of potatoes, a self-supporting A-frame storehouse can be constructed. A pit is dug about 10 feet deep and wooden air ducts are placed along the earthen floor. The roof of the building is constructed of wood, then covered with straw and soil.

Ducts for ventilation of bulk storage rooms can be laid out vertically as well as horizontally. The storeroom for potatoes provides for plenty of ventilation using simple materials. The room can be of any size or shape since air ducts can be positioned to extend evenly throughout.

When loading potatoes into bulk storage, even distribution of the produce is important for proper ventilation. Uneven loads will inhibit air movement and result in storage losses due to inadequate ventilation.

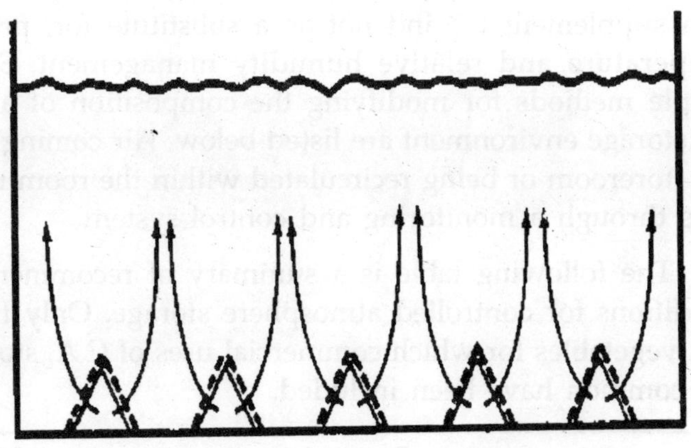

Even distribution of potatoes in the storeroom

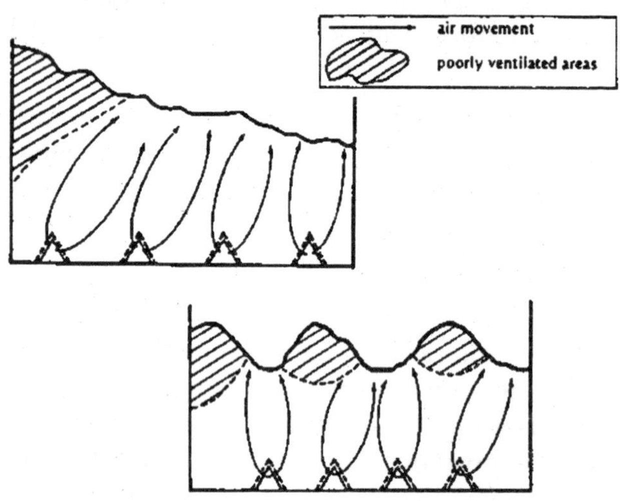

Uneven distribution of potatoes in the storeroom

Controlled or Modified Atmosphere (C.A.) Storage

Controlled or modified atmosphere storage should be used as a supplement to, and not as a substitute for, proper temperature and relative humidity management. Some simple methods for modifying the composition of air in the storage environment are listed below. Air coming into the storeroom or being recirculated within the room must pass through a monitoring and control system.

The following table is a summary of recommended conditions for controlled atmosphere storage. Only fruits and vegetables for which commercial uses of C.A. storage are common have been included.

	Temperature		$\%O_2$ range	$\%CO_2$ range
	C		F	
Strawberry	0-5	32-41	10	15-20

Apple	0-5	32-41	2-3	1-2
Kiwifruit	0-5	32-41	2	5
Nuts and dried fruits	0-25	32-77	0-1	0-100
Bananas	12-15	54-59	2-5	2-5
Cantaloupe	3-7	38-45	3-5	10-15
Lettuce	0-5	32-41	2-5	0
Tomatoes				
Mature green	12-20	54-68	3-5	0
Partially-ripe	8- 12	47-54	3-5	0

Controlled atmosphere storage of pallet loads of produce is also possible using a more permanent set-up for creating a gas-seal. Any number of pallets can be accomodated inside a plastic tent. A small trough constructed of sheet metal is laid in a rectangular pattern into a concrete floor of a storage structure. A 7-mil polyethylene sheet is put over the pallet load of produce, and the sheet is sealed by pushing a long piece of rubber tubing into the trough.

A low cost plastic tent fashioned from clear polyethylene sheeting can be used for controlled atmosphere storage of bunches of green bananas. A small fan serves to circulate the C.A. storage air (2% O2 and 5% CO2) through a chamber of potassium permanganate on aluminum oxide (Purafil). Ripening is delayed as ethylene is scrubbed from the storage air. The shelf life of bananas under these conditions is four to six weeks at ambient temperatures.

7

Packing Methods

In the simplest packinghouse, produce is delivered in picking containers, immediately after harvest, directly to the packers. The packers then sort, grade, size and pack the produce directly into appropriate transport containers. In this case, each worker must be knowledgeable regarding produce defects, grade and size requirements, and packing methods. As the size and complexity of the packinghouse increases, more operations and workers trained in specific tasks might be added.

Dumping

Produce must somehow be removed from the field bin or harvesting container and moved through the packinghouse. This first step is known as "dumping". Dumping must be done gently, whether using water assisted methods or dry dumping. Wet dumping can decrease bruising and abrasions by using moving, chlorinated (100-150 ppm) water to carry delicate produce. When using dry dumping, padded, sloped ramps or moving conveyor belts can decrease injuries to produce.

Pre-sorting produce is usually done to eliminate injured, decayed, or otherwise defective produce (culls) before cooling or additional handling. Pre-sorting will save energy in that culls will not be handled. Removing decaying produce items will limit the spread of infection to other units, especially if postharvest pesticides are not being used.

CLEANING

For some commodities, such as kiwifruits and avocadoes, dry brushing may be sufficient to clean the produce. Other commodities, however, such as bananas and carrots, require washing. The choice of brushing and/or washing will depend upon both the type of commodity and the type of contamination.

Wash before cooling and packing : tomatoes, cucumbers, leafy greens

Wash to remove latex, reduce staining: mangoes, bananas

Wash after storage : sweetpotatoes, potatoes, carrots

Dry brush after curing or storage: onions, garlic, kiwifruit

Do Not Wash : green beans, melons, cabbage, okra, peas, peppers, summer squash

Sanitation is essential, both to control the spread of disease from one item to another, and to limit spore buildup in wash water or in the packinghouse air. Chlorine treatments (100 to 150 ppm Cl) can be used in wash water to help control pathogen buildup during packing operations. There is some variation in the strength of bleach available commercially in different countries, but a rule of thumb is to use 1 to 2 mls of chlorine bleach per liter (1 to 2 ounces of chlorine bleach per 8 gallons of clean water). Walls, floors and packing equipment can also be cleaned using quaternary ammonium compounds labeled as safe for food processing equipment.

Waxing

Waxing of immature fruit vegetables such as cucumbers and summer squash; mature fruit vegetables such as eggplant, peppers and tomatoes; and fruits such as apples and peaches is common. Food grade waxes are used to replace some of the natural waxes removed in washing and cleaning operations, and can help reduce water loss during handling and marketing. If produce is waxed, the wax coating must be allowed to dry thoroughly before further handling.

Sizing

Sizing produce is optional but may be worthwhile if certain size grades receive a higher price than others. In most low-input packinghouses, manual sizing is still commonly practiced. Operators should be trained in selecting the size desired and to either directly pack the items into containers or place the selected produce gently into a bin for packing further down the line. Sizing can be done subjectively (visually) with the use of standard size gauges. Examples of the smallest and largest acceptable sizes for each product can be placed within view of the operator for easy reference. Hand held sizers are used for a variety of products.

Many products have established U.S. grades and standards that can assist packers in sorting and sizing produce. The following are examples of standards based upon diameter and/or length.

Several types of mechanical sizers are available for small scale operations. One type is composed of a long slanted tray with a series of openings which converge (largest at the top, smallest at the bottom). This type of sizer works best with round commodities. Other sizers are designed as conveyors fitted with chain or plastic belts with various sized openings, and are useful for sizing most commodities. Another simple method for mechanical sizing is to use a set of diverging bar rollers, where the smallest sized produce falls through the rollers first to a sorting belt or bin, and larger sized produce falls between successively more divergent rollers.

Narrow Pallet System

The University of Wisconsin is promoting this handling system for produce because it saves time, money and is safer for handlers. With a hand pallet truck you can move up to 16 cartons at a time. This system can cut your time spent moving boxes and will dramatically reduce the stress on your body.

A hand pallet truck is similar to a regular hand truck (dolly) except that the base uses pivoting forks instead of a fixed plate metal shoe. If you stack your load on a small plastic pallet (14 x 24 inches), you can position the forks underneath and move an entire stack of cartons at once. A hand pallet truck can be tilted (by releasing the forks) for loading and unloading and locked into an upright position to tip back and roll the load.

According to the UWisconsin project, custom-made hand pallet truck and pallet costs approximately $750. At $7.00 per hour, you'd need to save 107 hours for the system to pay for itself. If you saved 30 minutes per weekday (10 hours per month), the system would pay for itself in less than 11 months. The pallet truck system also saves money in reduced back pain and therefore fewer visits to the doctor, chiropractor or a massage therapist. If you saved yourself 15 visits (at $50/visit), the system would pay for itself.

General operations in a Packinghouse

The typical series of operations in a packinghouse are

illustrated below. Dumping can be done using either dry or water-assisted methods, depending upon the sort of produce being handled. Cleaning, as well, can be by washing with chlorinated water or dry brushing alone. Waxing, if practiced, occurs after washing and removal of surface moisture. Grading, as illustrated, separates the product into processing and fresh market categories. Sizing further separates the product, with the smallest size going to the local market or to processing. Typically, the best quality produce is packaged and marketed at the regional or national level.

Packing operations that are done in an unsystematic manner can cause delays, add costs or effect produce quality. You can save time and money by laying out the packing shed in an organized, step-by-step system.

Any time produce is dumped from one container into another, care should be taken to reduce mechanical damage to the commodity. When dumping produce from field bins or from transport vehicles into the packinghouse, dry or wet dumping can be practiced. When using dry dumping practices, the field container should be emptied slowly and gently onto a tilted ramp with padded edges. In the illustration below, a conveyor belt then carries the dry dumped produce into the packinghouse.

Wet dumping is sometimes used to reduce mechanical damage, either by dumping into water rather than onto a dry ramp, or by immersion and floatation. If the specific density of the produce, such as apples, is lower than that of water the produce will float. For some produce, such as pears, salts (such as sodium lignin sulfonate, sodium silicate or sodium sulfate) must be added to the water to increase its specific density and assure fruit floatation.

The canvas curtain illustrated below is used to break the fall of fruit moving from a conveyor into a bulk bin.

Conveyor equipment

To reduce bruising:

— use foam-padded ramps

— slow produce speed on ramps by installing retarding flaps, curtains, blankets or drapes

— use a flexible drape to slow fruit as it falls from a higher conveyor belt to a lower one (use a lightweight drape for short transition, heavier drape for higher transition)

— belting to prevent produce from falling directly onto roller sizers

— remove support pan from under belt in an area where fruit lands on a new belt.

Steel drums can be used to make a simple washing stand. The drums are cut in half, fitted with drain holes, and all the metal edges are covered with split rubber or plastic hose. The drums are then set into a sloped wooden table. The table top is constructed from wooden slats and is used as a drying rack before packing. Because steel drums are often used to store petroleum and chemical products, they should be thoroughly cleaned before being used as a washing stand.

This tank for washing produce is made from galvanised sheet metal. A baffle made of perforated sheet metal is positioned near the drain pipe and helps to circulate water through the produce. Fresh water is added under pressure through a perforated pipe, which helps move floating produce toward the drain end of the tank for removal after cleaning.

Improvements to the design might include a removable trash screen in front of the baffle, and/or a re-circulating system for the wash water (with the addition of chlorine). Chlorination of wash water is very important. Chlorine can reduce the spread of contamination form one item to another during the washing stage. The pH of the wash water should be maintained at 6.5 to 7.5 for best results.

Typically 1 to 2 mls of chlorine bleach per liter of clean water will provide 100 to 150 ppm total Cl. More chlorine will be required if temperatures are high or if there is a lot of organic matter in the wash water.

The waxing device illustrated here is designed to be used after a series of dry brushes on a conveyor line. Industrial wool felt is used to distribute the liquid wax to the fruits or vegetables from a trough made the same width as the belt. Evaporation of wax from the felt is reduced by covering the felt with a layer of heavy polyethylene sheeting.

Incoming produce is placed in the sorting bin, sorted by one worker into the packing bin, and finally packed by a second worker. If workers must stand to sort produce, a firm rubber pad for the floor can help reduce fatigue. The surface of the portable sorting table illustrated below is constructed from canvas and has a radius of about 1 meter (about 3 feet). The edges are lined with a thin layer of foam to protect produce from bruising during sorting, and the slope from the center toward the sorter is set at 10 degrees. Produce can be dumped onto the table from a harvesting container, then sorted by size, color and/or grade, and packed directly into shipping containers. Up to 4 sorters/packers can work comfortably side by side.

When sorting for rejects, and removing any product that is too small, decayed or damaged, the height of the sorting table should be set at a level comfortable for sorters. Stools, or a firm rubber pad on which to stand, can be provided to reduce fatigue. Locations of the table and the sorting bins should be chosen to minimise hand movements.

It is recommended that the workers' arms create a 45 degree angle when s/he reaches toward the table, and that the width of the table be less than 0.5 meter to reduce stretching. Good lighting (500 to 1000 lux at work surface) will enhance the ability of the sorter to spot defects, and dark, dull belts or table tops can reduce eye strain.

If a conveyor system is in use, the product must not flow too fast for the sorters to do their work. The rotational speed of push-bar or roller conveyors should be adjusted to rotate the product twice within the immediate field of view of the worker. Periodical rotation of worker positions on the line will help to reduce monotony and fatigue. Supervisors should be able to quickly identify under-sorting and over-sorting.

Sizing

Round produce units can be graded by using sizing rings. Rings can be fashioned from wood or purchased ready-made in a wide variety of sizes.

The onion sizing table is one of three (or more) tables used in a stairway fashion. Each table is made of plywood, and has been perforated with holes of a specific size. The uppermost table has the largest size holes, and the lowest table has the smallest holes. A layer of onions is dumped onto the uppermost table. Those that do not pass through are classified as "extra-large" in size. Those that pass through the holes in the table fall into a mesh bag and roll into a large container. This container of onions is dumped onto the second sizing table. The onions that do not pass through are classified as "large", and so on.

The citrus fruit sizer is composed of a rectangular chute made of plywood, padded with foam to prevent bruising. The fruit is dumped into the octagonal platform at the top of the chute, then allowed to roll, one by one, down toward a series of constrictions. Large fruits are caught in the first constriction, medium in the second, and small in the last. Undersized fruit passes out the end of the chute directly into a container. Workers must manually remove each fruit and place it into the appropriate size container before the next fruit can pass through the chute. The sizing is fastest when five workers are stationed at the sizer.

If a conveyor system is used in the packinghouse, a wide variety of sizing chains and belts are available for sorting produce. Sizing chains can be purchased in many widths and in any size opening. Square openings are usually used for commodities such as apples, tomatoes and onions, while rectangular openings are used for peaches and peppers. Hexagonal openings are often used for potatoes and onions.

Small scale equipment for packing produce is available from several manufacturers and suppliers. It is a fruit packing line, available from TEW Manufacturing Corporation at a cost of US$ 5000 to $6000, depending upon size. This particular model includes a receiving belt, washer, water absorber and sorting table. Optional equipment includes eliminator, dry brusher, waxer, and a variety of sizers, rolling inspection belts, conveyor belts and side tables.

Packing Materials

Throughout the entire handling system, packaging can be both an aid and a hindrance to obtaining maximum storage life and quality. Packages need to be vented yet be sturdy enough to prevent collapse. If produce is packed for ease of handling, waxed cartons, wooden crates or rigid plastic containers are preferable to bags or open baskets, since bags and baskets provide no protection to the produce when stacked.

Sometimes locally constructed containers can be strengthened or lined to provide added protection to produce. Waxed cartons, wooden crates and plastic containers, while more expensive, are cost effective when used for the domestic market. These containers are reusable and can stand up well to the high relative humidity found in the storage environment. Adding a simple cardboard liner to a crate will make it less likely to cause abrasion to produce.

Containers should not be filled either too loosely or too tightly for best results. Loose products may vibrate against others and cause bruising, while over-packing results in compression bruising. Shredded newspaper is inexpensive and a lightweight filler for shipping containers.

For small-scale handlers interested in constructing their own cartons from corrugated fiberboard, Broustead and New provide detailed information. Many types of agricultural fibers are suitable for paper making, and handlers may find it economically sensible to include these operations in their postharvest system. Corrugated fiberboard is manufactured in four flute types-type B is the most commonly used for handling perishables.

Whenever packages are handled in a high humidity environment, much of their strength is lost. Collapsed packages provide little or no protection, requiring the commodity inside to support all of the weight of the overhead load. Packing is meant to protect the commodity by immobilising and cushioning it, but temperature management can be made more difficult if packing materials block ventilation holes. Packing materials can act as vapour barriers and can help maintain higher relative humidity within the package. In addition to protection, packaging allows quick handling throughout distribution and marketing and can minimise impacts of rough handling.

Produce can be hand-packed to create an attractive pack, often using a fixed count of uniformly sized units. Packaging materials such as trays, cups, wraps, liners and pads may be added to help immobilise the produce. Simple mechanical packing systems often use the volume-fill method or tight-fill method, in which sorted produce is delivered into boxes, then vibration settled. Most volume-fillers are designed to use weight as an estimate of volume, and final adjustments are done by hand.

Ethylene absorber sachets placed into containers with ethylene sensitive produce can reduce the rate of ripening of fruits, de-greening of vegetables or floral wilting.

Packaging in plastic films can modify the atmosphere surrounding the produce. MAP generally restricts air movement, allowing the product's normal respiration processes to reduce oxygen content and increase carbon dioxide content of the air inside the package. An additional major benefit to the use of plastic films is the reduction of water loss.

MAP can be used within a shipping container and within consumer units. Atmospheric modification can be actively generated by creating a slight vacuum in a vapour sealed package (such as an un-vented polyethylene bag), and then replacing the package atmosphere with the desired gas mixture. In general, lowering oxygen and increasing carbon dioxide concentrations will be beneficial for most commodities. Selection of the best polymeric film for each commodity/package size combination depends upon film permeability and the respiration rate of the commodity under the expected time/temperature conditions during handling. Absorbers of oxygen, carbon dioxide and/or ethylene can be used within packages or containers to help maintain the desired atmospheric composition.

Modified atmosphere packaging should always be considered as a supplement to proper temperature and relative humidity management. The differences between beneficial and harmful concentrations of oxygen and carbon dioxide for each kind of produce are relatively small, so great care must be taken when using these technologies.

Packing Station Practices

When trimming is necessary, add a loose board, thick enough to reach above the height of the front rail. The front rail should be smooth and rounded. A simple field packing station can be constructed from wooden poles and a sheet of polyethylene. Thatch over the roof will provide shade and keep the station cool. The structure should be oriented so that the roof overhang keeps out the majority of the sun's rays.

Hands of bananas, after undergoing washing to remove latex and perhaps spraying with fungicides, are typically packed into cardboard containers lined with polyethylene. The following illustrations depict one method of filling a container with the fruit in order to ensure less damage during transport. Note that the polyethylene liner is folded up over the bananas before closing the box.

— Wide, flat medium-to-small hand in middle of compartment

— Medium-length, wide hand on top, crown not touching fruit below

— Medium-to-short length, wide hand, crown not touching fruit below

— One large hand, or two clusters with long fingers

A circular rotating table can be used to pack a variety of crops. The produce is fed in along a conveyor or if no conveyor is in use, simply put onto the table, where packers select the produce and fill cartons at their stations. In the illustration below, a discard belt has been added below the supply belt, allowing easy disposal of culls.

Each packer can work independently, trimming as needed and check weighing cartons on occasion.

Containers

Sacks are often used to package produce, since they tend to be inexpensive and readily available. The following table provides some information regarding the characteristics of different kinds of materials used to make sacks. None of the types of sacks available are good for protecting fresh produce, and they should be avoided whenever possible.

Containers can be constructed from wood and wire. A special closing tool makes bending the wire loops on the crate's lid easier for packers to do. Wire-bound crates are used for many commodities including melons, beans, eggplant, greens, peppers, squash and citrus fruits. Package Research Laboratory can provide a list of suppliers in your area.

A wooden lug is the typical packing container for table grapes. This container is very sturdy and maintains its stacking strength over long periods of time at high relative humidity. Rigid plastic containers are also widely used.

Often, a paper liner is folded over the grapes before the top is nailed closed. The liner protects the produce from dust and water condensation. If a pad containing sulfur dioxide can be enclosed with the grapes within a plastic liner as a treatment to control decay. Most commodities other than table grapes can be damaged (bleached) by sulfur dioxide treatments.

Rigid plastic or wooden containers are also used extensively for asparagus. The trimmed spears are packed upright in containers that provide for a large amount of ventilation. For domestic marketing, plastic crates provide excellent protection for produce and adequate ventilation during handling, cooling, transport and storage. Some plastic crates are collapsible or can be nested when stacked for easier handling when empty. Crates should be cleaned on a regular basis with chlorinated water and detergent to reduce the chances of spreading decay from one load to the next.

Containers for cut flowers are often long and narrow, of full telescopic design with vents at both ends to facilitate forced-air cooling. The total vent area should be 5% of the total box surface area. A closable flap can help maintain cool temperatures if boxes are temporarily delayed in transport or storage in an uncontrolled temperature environment. A simple wooden tray with raised corners is stackable and allows plenty of ventilation for fragile crops such as ripe tomatoes.

Choosing a container can often be a complex decision. There will always be trade-offs to consider with respect to price, durability, reusability and protection. Here is a Checklist of factors to consider when selecting a container from two or more package types. Use as many copies of this table as are required to compare all your packaging options.

Describe each package type, then fill out the table to see which package has the most positive characteristics and will therefore be the best choice for the intended market.

— Give ratings to the characteristics for different types of packaging.

+2 = excellent

+1 = good

0 = fair

-1 = unsuitable

-2 = extremely unsuitable.

— Give each characteristic a value from 0 to 100. The total for all 24 characteristics for each package should add up to 100.

— Multiply for each package the rating for each characteristic times its value and record the new value.

— Add up the total points for each type of package. Does the package that receives the most points match your preference? What are the differences between the two packages with the highest score? If one is lower in price that may be your best choice.

Packaging Practices

Adding a fiberboard divider to a carton will increase stacking strength. The use of dividers is common with heavy crops such as melons. The dividers also prevent melons from vibrating against one another during handling and transport. Wooden inserts, or fiberboard folded into triangles and placed in all four corners can be especially useful when a carton needs strengthening. When locally made containers have sharp edges or rough inner surfaces, a simple, inexpensive inner made from fiberboard can be used to protect produce from damage during handling.

If large bags or baskets must be used for bulk packaging of fruits or vegetables, the use of a simple vent can help reduce the buildup of heat as the product respires. In the illustration below, a tube of woven bamboo (about one meter long) is used to vent a large bag of chili peppers. Place the vent tube into the container before filling the container with product.

The woven bamboo tube allows air to reach the center of the sack or basket during handling. Thin paper or plastic sleeves are a useful material for protecting cut flowers from damage during handling and transport. A packer is pulling a sleeve up over a bunch of flowers before packing the flowers into a vented fiberboard carton. The sleeves both provide protection and help keep the bunches of flowers separate inside the box.

Labelling Packages

Labelling packages helps handlers to keep track of the produce as it moves through the postharvest system, and assists wholesalers and retailers in using proper practices. Labels can be preprinted on fiberboard boxes, or glued, stamped or stenciled on to containers. Brand labelling packages can aid in advertising for the product's producer, packer and/or shippers. Some shippers also provide brochures detailing storage methods or recipes for consumers.

Shipping labels can contain some or all of the following information:

— Common name of the product.
— Net weight, count and/or volume.
— Brand name.
— Name and address of packer or shipper.
— Country or region of origin.
— Size and grade.
— Recommended storage temperature.
— Special handling instructions.
— Names of approved waxes and/or pesticides used on the product.

Labelling of consumer packages is mandatory under FDA regulations. Labels must contain the name of the product, net weight, and name and address of the producer, packer or distributor.

Modularisation

When a variety of different sized cartons are packed at the same time, using boxes in standard sizes can greatly ease future handling. When handling boxes that are non-uniform, stacks can be unstable or heavier cartons can crush lighter ones. An unstable load is likely to fall over during transport or to collapse during storage.

Recommended container sizes are shown below. These containers are part of the MUM programme (Modularisation, Unitisation and Metrication) advocated by the USDA. They can all be stacked in a variety of patterns, depending upon their size, yet still form a stable load on a single pallet of 1000 x 1200 mm (40 x 48 inches).

The following illustrations show the arrangement of the most popular MUM containers on a standard pallet (1000 x 1200 mm or 40 x 48 inches). Using MUM containers can save space during transport and storage, since pallet utilisation is close to 100%.

Modified Atmosphere Packaging (MAP)

Within a consumer package: If commodity and film permeability characteristics are properly matched, an appropriate atmosphere can evolve passively through consumption of CO_2 and production of CO_2 during respiration. Some rigid plastic consumer packages are designed with a gas diffusion window.

Fresh-cut or lightly processed lettuce (shredded or chopped) can be packaged in 5-mil plastic bags. After a partial vacuum is created, a gas mixture of 30 to 50% CO_2 and 4 to 6% CO_2 is introduced into the bag, which is then sealed.

SEALED PLASTIC BAG

Within a shipping container: Polyethylene liners are added to shipping containers in cherry boxes, and polyethylene bags are used for bananas destined for distant markets.

Within a pallet: A single pallet load of produce such as strawberries can be sealed within a shroud of 5 mil polyethylene bag and a plastic sheet on the pallet base using wide tape. A slight vacuum can be introduced and 15% CO_2 added to the air introduced via a small hose.

Many plastic films are available for packaging, but very few have gas permeabilities that make them suitable for MAP. Low density polyethylene and polyvinyl chloride are the main films used in packaging fresh fruits and vegetables. Saran and polyester have such low gas permeabilities that they are suitable only for commodities with very low respiration rates. The following table provides the permeabilities of the films currently available for packaging fresh produce.

Many shippers and receivers prefer to handle unit loads of produce pallets rather than handling individual shipping containers. The switch to unit loads has reduced handling, causes less damage to the containers and produce inside, and allows faster loading/ unloading of transportation vehicles. If small scale handlers wish to use unit loads for shipping produce, either wooden pallets or slip sheets can serve as the base of the load. Using guides for aligning the boxes (such as placing the pallet to be loaded against the corner of a room, or building a set of "bounce boards" if the pallet is loaded outside) will stabilise the load. Wooden pallets must be made perfectly square for the pallet load to be as stable as possible.

Using fiberboard, plastic or wooden containers with vertical interlocking tabs can also help improve the unit load's stability. Containers must have holes for ventilation which align when stacked squarely on top of one another. Glue can be used between layers of containers to reduce slipping, and plastic netting or plastic or metal straps should be added to secure the load. Corner-boards made from fiberboard, plastic or metal help to provide for a stable unit load.

8

Transportaion of Horticultural Products

Mixed loads can be a serious concern when temperature optima are not compatible (for example, when transporting chilling sensitive fruits with commodities that require very low temperatures) or when ethylene producing commodities and ethylene sensitive commodities are transported together. High ethylene producers (such as ripe bananas, apples, cantaloupe) can induce physiological disorders and/or undesirable changes in color, flavour and texture in ethylene sensitive commodities (such as lettuce, cucumbers, carrots, potatoes, sweet potatoes).

A wide range of pallet covers are available for covering cooled product during handling and transportation. Polyethylene covers are inexpensive and lightweight, and protect pallet loads from dust, moisture and some loss of cold. Lightweight insulated covers can protect the load from heat gain for several hours (for example, during a delay in loading). Heavyweight covers are sometimes used to protect tropical products from the cold when shipped during winter. Visit the internet for more information (keywords: produce pallet covers) or order a catalog from the International Ripening Company.

VEHICLES

Bulk loads of produce should be carefully loaded so as not to cause mechanical damage. Vehicles can be padded or lined with a thick layer of straw. Woven mats or sacks can be used in the beds of small vehicles. Other loads should not be placed on top of the bulk commodity.

Cooling open loads is desirable whenever possible. A truck ventilating device can be constructed for an unrefrigerated open vehicle by covering the load loosely with canvas and fashioning a wind catcher from sheet metal. The scoop should be mounted at the front of the bed and should reach somewhat higher than the height of the cab. High transportation speeds and/or long distance transport run the risk of causing excess drying of the crop.

This ventilating system was designed for hauling bulk loads of fresh fava beans in Iran. The wind catcher and ducts were constructed using wooden crates. After removing their end panels the crates were wired together into the pattern shown below. Air flows upward through the load during transport, helping to keep the produce from overheating. This system has also been used in pick-up trucks, and for hauling bulk greens and green beans. Best results are obtained when transporting during the early morning hours, before sunrise.

USDA Post Harvest Cooling Programme

The USDA post harvest cooling programme stresses to small growers the need to:

— sort and grade produce out of the field,

— package the produce properly for the market,

— immediately cool the produce to remove field heat.

This small cooler uses a 12,000 BTU/hr (1 ton) 110 volt

room window air conditioner to cool air inside the insulated box. The cool air inside the front of the box is forced through the produce by a pressure fan in a second wall inside. The return air passes under a false floor to the front of the box.

The Portacooler was built by a cooperating team at USDA, Beltsville, Maryland, to cool highly perishable berries and other air cooled produce. Approximate cost is $US 1200, but the cost can be considerably lower if it is built using a used air conditioning unit.

Refrigerated Trailers

For optimum transport temperature management, refrigerated trailers need insulation, a high capacity refrigeration unit and fan, and an air delivery duct. The checklist below includes these and other desired features in a top-air delivery trailer.

The condition of the inside of a refrigerated trailer affects its ability to maintain desired temperatures during transport. Handlers should inspect the trailer before loading, and check these features:

Stacking Patterns

Bushels (very sturdy, uniformly made, stackable baskets) of produce can be loaded into a refrigerated trailer using a pattern of alternately inverted layers that leave plenty of space between rows for air circulation.

Produce transported in cartons should also be stacked so as to allow adequate air circulation throughout the load. The diagram below illustrates crosswise offset loading of partial telescopic containers. On the floor of the truck, pallets or other supports should be used to keep the cartons out of direct contact with the floor.

When cartons of various sizes must be loaded together, the larger, heavier containers should be placed on the bottom of the load. Parallel channels should be left for air to move through the length of the load.

Often the large containers used for cut flower packaging must be hand-stacked when loaded into a transport vehicle. The best loading pattern for cut flowers is known as the pigeon hole pattern, where boxes are stacked in alternating solid and open layers, and channels are left down both side walls. This pattern provides channels for air circulation lengthwise through the load, and allows every box to be in direct contact with refrigerated air.

Containers should be loaded so that they are away from the side walls and the floor of the transport vehicle in order to minimise the conduction of heat from the outside environment. In the diagrams below, the numbers of cartons refer to how many cartons would be in contact with the walls and floor of the truck when fully loaded.

Only the load on the bottom right is fully protected from heat transfer. The use of pallets keeps the cartons off the floor, while center-loading leaves an insulating air space between the pallet loads and the outside walls.

When stacking containers, be sure to align them properly. Whenever possible, stack them so that corner matches corner on both the cartons and the pallet. Most of the strength of corrugated fiberboard containers is in their corners, so an overhang of only 1 inch will decrease stacking strength by 15 to 34%.

There should always be a void between the last stack of produce and the back of the transport vehicle. The load should be braced to prevent shifting against the rear door during transit. If the load shifts, it can block air circulation, and fallen cartons can present great danger to workers who open the door at a destination market. A simple wooden brace can be constructed and installed to prevent damage during transport.

Bracing the load can be accomplished using wooden braces, load rails, air pillows or Styrofoam blocks. The key is to immobilise stacked produce to reduce damage during transport.

Air Transport

To prevent shifting of the load in a cargo container for air transport, a piece of solid foam or folded fiberboard should be placed along the curved or triangular portion of the floor of the container. Cartons stacked on top will be much better supported and be held upright.

Handling Produce at its Destination

When handling produce at its destination, again it is important to avoid rough handling, minimise the number of handling steps, and maintain the lowest feasible temperature. If produce is to be stored before sale, then wholesale and retail markets need clean, well insulated storage rooms. Since a variety of commodities is usually being handled simultaneously at this point, it is important to remember not to mix those with different temperature requirements or store ethylene sensitive commodities near ethylene generating commodities. Stacking of non-uniform containers should be done with care to prevent collapse of weaker packages.

Before produce is sold to the consumer, the handler may wish to sort for quality, or at least to discard any damaged or decayed produce in order to give the product more market appeal. If ripeness or maturity is non-uniform, sorting at destination can provide the seller with a higher price for the better quality produce.

If the produce handled is a climacteric fruit crop that was harvested before it was ripe (bananas, tomatoes, avocadoes, mangoes), the handler at destination may want to ripen the produce before it is sold to the public. Sometimes commodities such as bananas are left at ambient temperatures and allowed to ripen naturally. Covering the bananas with a plastic sheet will help ripening be more uniform throughout the lot. Placing a simple air vent (a pipe or a tube of some sort) into the center of the pile of ripening fruit can help reduce overheating during ripening and increase subsequent shelf life. The introduction of ethylene gas or ethylene-releasing compounds into a special storage environment (known as a ripening room or cabinet, depending upon size) is a more effective way to ensure uniform ripening.

Temperatures of the display tables or refrigerated supermarket displays should be suited to the commodity on sale. For example, while peppers and tomatoes look pleasing when displayed with lettuce, peppers and tomatoes are chilling sensitive, while lettuce is not. Misting commodities that can tolerate surface water (lettuce, broccoli, green onions) with cool clean water can help maintain a high relative humidity around the product.

Outdoor marketplaces suffer from a lack of temperature control and high air circulation, which can lead to desiccation of crops, which will be seen as shriveling and wilting. These marketplaces can often benefit by the increased use of shading and protection from prevailing winds.

Finally, the handler at destination can help reduce losses in the future by maintaining good records of the sources of losses suffered at the wholesale or retail level. Identifying whether losses were due to mechanical damage, decay/disease, immaturity or over-ripeness allows the handler to provide better quality feedback to produce suppliers.

Unloading Transport Vehicles

A loading dock can ease the work associated with handling horticultural produce at destination. Containers can be transferred more rapidly and with less bending and lifting. For large trucks, a loading dock 117 to 122 cm high (46 to 48 inches) functions well, while for small trucks or pickups a height of 66 to 81 cm (26 to 32 inches) is recommended.

A simple device for easing the work of unloading transport vehicles can be constructed using a two pulleys and a strong rope. One pulley can be mounted inside the truck at the front of the bed, the second outside on a portable post or on a stationary object like the side of a building. Containers can be hung directly by their handles or placed into a sling.

A simple set of stairs can be constructed to ease the work of loading and unloading produce. The stairs illustrated here can be folded and pushed back under the back of the truck when the vehicle is in motion. The steps can be made of wood or steel matting and steel bars can be used for supports.

Using a ramp is a simple and safe method for unloading produce. The ramp should be wide enough to prevent accidents and strong enough to carry the full weight of the handler plus the package she/he is carrying. Providing hand-trucks or small carts can also ease the work associated with unloading.

Suitable Temperatures

When produce is held at destination for a short time before marketing, the handler can help maintain quality and reduce losses by storing commodities at their most suitable temperature. However, if the storage period is seven days or less, relative humidity is maintained between 85 and 95%, and the ethylene level is kept below 1 ppm, by ventilating or using a scrubber, most commodities can be grouped into the following three temperature-based groups.

Sorting

Some produce may require washing, trimming, bunching or sorting at the wholesale or retail market level. The layout of the work station used for handling produce at destination should be organised to minimise nonproductive movement. In the illustration below, a dump table is located next to a sink for washing produce, and the drain board is positioned directly next to the sink. Once produce has dried, cartons can be packed and placed onto a cart located right next to the repacking table. With this layout, a single operator could easily perform all the handling steps or several handlers could work side by side.

Some produce may have to be repacked by the wholesaler or retailer due to changes in quality or uneven ripening. The tomato sorting table illustrated below has work stations for up to five who select either ripes, pinks or breakers and allow the green tomatoes to run off to the end of the line. Rejects (culls) are placed in pails under the table.

Ripening Process

Ripening is the process by which fruits attain their desirable flavour, color and textural properties. Climactericfruits can ripen off the plant once they have reached physiological maturity.

Climacteric fruits include apples, avocado, banana, blueberries, breadfruit, cherimoya, durian, feijoa, fig, guava, kiwifruit, mango, muskmelon, papaya, passion fruit, pears, persimmon, plantain, quince, sapodilla, sapote, soursop, stone fruits (apricots, nectarines, peaches, plums) and tomato. Some of these fruits if harvested "mature-green", can be ripened after harvest and short term storage. Pears and bananas are unusual in that they develop the best flavour and texture characteristics when harvested mature-green and ripened off the tree. Avocadoes do not ripen on the tree.

Some climacteric fruits give off large quantities of ethylene during ripening. These include apples, apricots, avocadoes, cantaloupe, kiwifruit, nectarines, peaches, pears, plums and passion fruit. A small dose of ethylene gas will stimulate other climacteric fruits to begin the ripening process. A few climacteric fruits, such as muskmelons, will not increase in sugar content during ripening, but will soften.

Non-climacteric fruits must ripen on the plant if you want a fully ripe fruit, since once they have been harvested, no further ripening will occur. Flavour and texture will be of low quality if fruits are picked before fully ripe.

Some non-climacteric fruits include berries, cherries, citrus fruits (lemons, limes, oranges, grapefruits, mandarins, tangerines), cucumber, dates, eggplant, grapes, lychee, okra, peas, peppers, pineapple, pomegranates, strawberry, summer squash, tamarillo and watermelon.

Non-climacteric fruits will not respond to attempts to ripen them with ethylene gas. A partially red strawberry, for example, will not develop any more color or sweetness after being picked, and will deteriorate faster if exposed to ethylene. Watermelons develop most of their sweetness during the week before they reach full maturity, making early harvest very undesirable.

Sometimes ripening commodities before sale at the wholesale or retail level will improve their value. Ripening rooms are often used for tomatoes, citrus fruits and bananas. The use of diluted ethylene gas mixtures is safer than using pure ethylene which is explosive and flammable at concentrations of 3% or higher.

For tomatoes, technical grade ethylene gas is introduced into the room at a concentration of about 100 ppm for about 48 hours. Approximately 0.25 cubic feet/ hr of ethylene gas is required for each 1000 cubic feet of ripening room volume. A small fan can be used to ensure a uniform continuous flow of ethylene into and through the room. Forced-air ripening is increasingly being used to provide more uniform temperatures and ethylene concentrations throughout the ripening room.

Continuous Flow Gassing:

Ripening can also be initiated by using ethylene generated by passing ethanol over a bed of activated alumina. This method is safer than using pure ethylene gas. Ethylene-releasing compounds such as ethephon {(2-chloroethyl) phosphoric acid} are sometimes used to ripen tomatoes destined for processing. When using ethephon as a spray, the amount of ethylene released will increase as pH and/ or relative humidity increase.

The following illustration is a de-greening room designed for use with citrus in pallet boxes. The ceiling of the room is relatively high, allowing boxes to be stacked at least four high. A false ceiling is added to provide for adequate air movement throughout the room. For more detailed information on room construction, temperature and relative humidity management and air circulation, refer to the article entitled "Ripening Facilities" in the reference section of the manual.

Several small ripening rooms may be more useful than a single large room for small scale handlers, since the amount of product handled at destination may vary from time to time. In this case, flow through systems can be designed to allow the use of one or more rooms at the same time. Flowmeters can be located in one place for ease of monitoring, or can be strung out in a line. Locating all the flowmeters in one place requires the use of more tubing than if flowmeters are located in each ripening room.

Small-scale handlers can now lease portable ripening facilities from a variety of companies in the United States. The self-contained, portable system illustrated below features a 20 pallet capacity, high capacity air flow and is simple to operate. All that is needed is a loading dock and a supply of 220V electricity.

Small-scale wholesalers and retailers can ripen fruits in bins or large cartons by placing a small quantity of ethylene-generating produce such as ripe bananas in with the produce to be ripened. Cover the bin or carton with a plastic sheet for 24 hours, then remove the plastic cover.

A simple way to ripen fruits at home in small amounts is to use a ripening bowl. Fruits that require ripening should be placed into the bowl with a ripe apple or ripe banana (or any other high ethylene-generating product). The bowl shown below is made of clear molded plastic and has ventilation holes around the top. Using this method, ripening will take from one to four days. Home ripening is also possible using another, extremely low-tech practice-- place fruits to be ripened into a paper bag with a ripe piece of fruit, close loosely and check in a few days.

Wooden Display Table

This wooden display table is designed to be used for commodities such as cruciferous crops or leafy green vegetables that tolerate cooling with ice. The table can be used in the horisontal position or as a tilted display. For more complete design specifications, contact the Cornell University Extension Service, 304 Riley-Robb Hall, Ithaca, New York, 14853.

Four to five lbs of crushed ice per square foot of display space are required for cooling per day. A catch pail should be provided for melt water. To minimise ice needs, the display tray should be insulated and kept out of the direct sun. When displaying horticultural crops, single or double layers of produce are most likely to protect the commodities from compression damage and over-handling by the consumers.

High relative humidity can be maintained during display by misting leafy vegetables and water tolerant crops with clean, cold water. A simple sprinkler device can be constructed by perforating a pipe with tiny holes and connecting it to a hose. If this display is used outdoors, shade should be provided. Displays and storage areas must be cleaned and sanitised on a regular basis. Trimmings, waste, and bruised product remaining in displays are unsightly and can be sources of decay, odour and ethylene.

9

Horticultural Marketing

Most farmers see themselves as "price takers", thinking that they have no control over prices and have to accept what is offered. They do not always know how to find new buyers nor how market demand is changing and which products are most profitable to grow. They lack the understanding to improve the prices they receive and the profitability of their production.

Your role as an extension advisor should be to help farmers become better informed about the market. They can then start to make decisions on how to improve their marketing. However, you must never tell farmers what they should do or what products to sell. Commercial decisions like these must be made by the farmers themselves. There are two reasons for this:

— Farmers need to become owners of new ideas. In this way they become committed and aware that they are responsible for the success or failure of what they do. This increases the chances of a successful outcome.

— If an extension officer (or NGO specialist) recommends a course of action that fails, he or she will be blamed by the farmers.

Extension workers should advise farmers of their different options, but they should never tell them what to do. If things go wrong, farmers will be very angry. When asked about their problems, farmers commonly identify marketing issues as their key constraint. Problems highlighted are usually lack of markets, poor prices, inadequate roads and poor communications.

However, while farmers can usually state clearly their problems they often face difficulties in identifying potential solutions. This guide aims to help you to help farmers work out what to do. Farmers are generally highly skilled in agricultural techniques but marketing requires learning new skills, new techniques and new sources of information. Armed with business and marketing skills farmers will be better able to run their farms profitably.

Small-scale farmers face the biggest marketing problems. They need the most support and that their success depends on getting the best prices possible. This can be done by obtaining better information about marketing and the different marketing options available to them.

Role of Marketing

Rural Economy

Those who carry out marketing have a strong incentive to increase the value of rural trade, because increased sales should lead to higher profits. Rural businesses include suppliers of inputs, buyers of produce, transporters, storage companies, processors and wholesalers. They can range in size from individual entrepreneurs to large-scale agribusinesses, but whatever their size, all stand to gain from improvements in the marketing process.

Businesses are often said to be exploiting farmers and making unfair profits. They certainly try to maximise their profits, but without such businesses farmers would not be linked to markets and would not be able to sell all their produce. Traders and other rural businesses can help farmers find new markets and lower their costs. All of this leads to improved production opportunities and higher incomes for farmers.

Rural marketing businesses are often small, have limited resources and are traditional in outlook. Identifying new markets, advising on technologies and improving understanding of markets are all ways in which extension workers can help traders and farmers to increase sales.

Governments can help farmers in many ways, without actually working with them directly. Promotion of competition, provision of market information and improvement of market infrastructure are powerful ways to ensure good returns for farmers.

Consumers

Consumers want to pay low prices. Farmers want to receive high prices and to be paid as much of the consumer price as possible. The best way of achieving a balance between these two conflicting aims is through an efficient and low-cost marketing chain. This generally involves using larger scale transport (achieving economies of scale), reducing losses, and reducing other costs. An example of how costs can be reduced is the promotion of competition between buyers.

Consumer preferences for food products are constantly changing and developing, particularly in the case of horticultural products. Consumers need a production-marketing chain that can respond to their changing tastes. The marketing system needs to deliver the volumes, quality and variety of safe and nutritious food products that consumers require. It needs to be sufficiently dynamic so that it can continue to provide consumers with choice by developing and delivering new products.

Effect of Profit

In the past, ministries of agriculture and even NGOs worked mainly on improving agricultural production and yields. This involved carrying out field trials, testing varieties and developing improved production techniques. The extension officer's primary role was to advise and encourage farmers to use improved production technologies.

Although farmers do, of course, benefit from improved yields, their primary concern must be to make a reasonable living from their farm. They need to generate sufficient profit from their sales to be able to cover their living costs. The key issue for them is how to do this.

The main role of the extension officers should therefore be to help farmers farm more profitably. To do this they must have a basic understanding of the factors that influence profits. These are farm gate or market prices, volumes sold and costs.

Profit Prices

Risk and Profit

High profits are closely linked to high risks. In agriculture, the enterprises that offer the possibility of the highest profit are usually the most risky. For example, if there is only a small demand for some products, such as flowers or some fruits, they can easily be oversupplied, leading to dramatic falls in price.

Some products may be technically difficult to produce or susceptible to pests and diseases. Farmers, and particularly poorer farmers, need to consider the importance of balancing production of potentially more profitable, but riskier, crops with those that provide a lower, but more stable and reliable income.

Enterprises that usually provide reliable incomes include staple crops and contract production for agricultural processors.

Prices of Farmer

The prices that farmers receive are mostly influenced by supply and demand.

Other important factors affecting the price a farmer receives include:

— *How much competition there is between buyers.* When there is only one buyer, he or she will possibly offer a low "take it or leave it" price. Prices are likely to be higher and more profitable when there are many buyers competing with each other.

— *The amount of information the farmer has.* A farmer who is poorly informed of market prices and demand will be less able to negotiate with traders.

— *The quality of the produce.* Buyers will sometimes, but not always, offer higher prices to producers with better quality products. If most farmers produce low quality produce it is difficult for a trader to market high-quality produce separately.

— *The transport costs.* Lower prices are generally offered to producers whose product is costly to transport.

Production- Marketing Chain

As produce moves along the production-marketing chain, prices increase. Unit prices are lowest when farmers sell a standing crop. An example of this is when fruit growers sell their fruit on the tree to fruit contractors, who then harvest and pack it. This is common in Near East and South Asian countries.

Harvested produce sold at the farm gate to rural buyers obtains a lower price than produce sold at a local assembly market which, in turn, is sold at a lower price than produce sold at a wholesale market. Retail sales achieve the highest price.

The price the farmer gets depends on the point in the marketing chain at which he or she decides to sell. Although prices are higher, selling further along the marketing chain involves additional costs for transport, market fees, meals and accommodation. There are also costs in terms of farmers' time. Farmers who take produce to market and sell directly to consumers will usually get the highest price but they do need to decide whether this is the best use of their time, as it may be more usefully spent managing the farm.

Identifying New Products and New Markets

Products sold in larger volumes, such as grains, may make low prices, but these prices are usually more stable. Such crops offer greater security but lower potential profits.

Products with an increasing demand offer good opportunities. The market is less likely to be oversupplied. Such products provide the opportunity for producers to expand their business as the market grows. The market for food in developing countries very often follows the same patterns of growth as have been experienced by developed countries. Examples are:

— urban consumers increasingly buy semi-prepared food (e.g. pre-packed salads or pre-cut vegetables) and processed food products;

- eating at fast-food restaurants becomes more popular and such restaurants become an important market for horticultural produce;
- snack products are increasingly eaten;
- retailing becomes more sophisticated and supermarkets start to take a larger share of sales.

New products and new markets provide opportunities for more rapid growth and more profitable production. For example, processors and supermarkets sometimes like to secure supply of crops by contracting farmers or intermediaries to supply them. These changes in the market for food lead to the development of new marketing channels, new buyers and a move away from more traditional assembly and wholesale markets.

Farmers' Marketing Costs and the Trader

Marketing costs can be reduced, especially by achieving economies of scale. The larger the volume of product marketed, the greater the scope for lowering unit costs, for example through using larger vehicles or bulk-buying of packaging. Sometimes, traders operating on a commission basic will accept lower percentage commissions for large quantities. Small farmers working on their own are naturally at a disadvantage. Cost savings have sometimes been achieved by farmers working as a goup to jointly purchase inputs and hire transport.

Traders also have costs. Because these are not always visible, traders are often accused of making unreasonable profits. Although traders can sometimes exploit a situation and make exceptional profits, there are also occasions when they are unable to sell products that they have purchased and therefore make losses. High profits on some occasions compensate for risks of losing money on others.

Losses of Products

Losses or products will inevitably occur during marketing and are one example of a marketing cost that is often invisible. Types of loss are:

Loss of weight. Produce can lose weight, mainly through water loss in storage, transport and marketing. A trader may buy a tonne of products, but as a result of water loss may only sell 975 kilos.

Loss of value over time. The fresher the product, the higher the price. Fresh produce will last longer, will probably taste better and may be more nutritious. Over time, bruises and pest damage become more apparent and water loss leads to many products starting to shrivel, all leading to losses in quality and in price. The most obvious loss occurs when produce has to be thrown away.

Physical losses. Fresh produce losses in developing countries can be exceptionally high. Losses of ten to thirty percent are often quoted. In practice, only on rare occasions is this volume of produce thrown away. These figures more accurately reflect the loss in value that occurs from the combination of moisture loss, damage, disease and loss of freshness, as well as the proportion of the harvest that is thrown away.

For a trader, minimising losses is very important. Part of the difference between the consumer price and the price paid to the farmer is because the trader needs to make a large margin, to cover wasted or damaged produce. By only looking at the selling price it is easy to think that the trader is making an unfair profit. However, when losses are taken into account the picture changes.

Factors Affecting Supply and Demand

The quantity of produce that consumers want to purchase

is affected by many factors, the most important being:

— price of the goods;

— tastes and preferences of the consumers;

— number of consumers;

— incomes of consumers;

— prices of competing produce;

— range of products available to consumers.

The quantity that producers supply is also affected by a number of factors, the most important being:

— price of the goods/products on the market;

— price of inputs/costs of production;

— technological factors;

— climate;

— storage possibilities.

The price of a product is mainly determined by supply and demand. Basically, a balance is achieved between what people are prepared to supply at a price and what people are willing to pay for the product. As the price rises the quantity that will be supplied also rises and the quantity demanded falls, and vice versa. It is important to note that:

Supply is what producers are prepared to sell at a certain price Demand is how much consumers are prepared to buy at the market price

While supply is influenced by production it is not always the same as production. For less perishable crops, farmers or traders may decide to store them in the hope that prices will rise, rather than sell them immediately. When prices do rise they may take the products out of store to sell. At this time supply is equal to production harvested for immediate sale plus products taken out of store.

However, it should be stressed that demand is not how much people would like to buy, nor what they should buy for a healthy lifestyle. It is what they are prepared to buy at the prevailing price.

Short-term, Seasonal and Long-term Price Changes

Unlike the prices of staple foods such as maize or rice, horticultural product prices fluctuate enormously. They can fluctuate from day to day and during the day, depending on supply and demand.

The main causes of short-term price changes of fresh produce are:

- The amount of produce on sale in the market on a particular day and the quantities sold in the previous few days.
- Short-term demand changes (e.g. holidays and festivals).
- The effect on demand of prices of competing products.

To take advantage of opportunities when prices are high a supplier needs to be in close communication with the markets and able to transport produce rapidly. In many countries farmers are increasingly able to deliver products, such as leafy vegetables, to markets when prices are high, using mobile telephones to contact buyers.

In countries with pronounced seasons, supplies are low at the start of the harvest season, so prices are high. Prices are at their lowest when the crop reaches maturity in the main production areas. At the end of the season prices normally increase again as supply diminishes. Prices are generally highest during the off-season, when only a small percentage of farmers are able to grow the crop.

Farmers who are located in areas where early or late-season crop production is possible (for example, hill or mountain areas) or who can use production methods, such as plastic tunnels or greenhouses, that bring forward the harvesting date, are best placed to take advantage of high early or late-season prices.

Similarly, production under irrigation can supply crops in the off-season when prices are normally the highest.

The theoretical point at which supply and demand are in balance is referred to by economists as "equilibrium". At this price there is enough incentive for farmers to produce the quantity that consumers will buy at that price. In practice, although the marketing chain tries to achieve this balance it is rarely done because there are so many factors affecting both supply and demand and because farmers lack adequate information about demand.

Understanding of Horticultural Markets

The way horticultural markets operate is complex. Markets are not very rational. Very often they appear to panic or overreact. If traders believe that there is a shortage of a product, prices will rise. Often, the increase is out of all proportion to any shortfall in supply. The converse is also true. If the market expects even a small oversupply, then prices fall rapidly.

Supply of Changes in Price

High prices have a large effect on farmers' profits. In the short term a farmer's response to high prices will be to try to increase the quantity of produce marketed. For example, tomato farmers may harvest green, unripe tomatoes. In the longer term the farmer will consider expanding the crop area and look for ways to increase production. In response to high prices farmers usually increase their production in the next season. As many of them make the same decision, there is a large increase in supply and oversupply occurs. Prices fall and farmers reduce production in the following season.

Demand of Changes in Price

When prices fall consumers increase their purchases. They buy less when prices rise. However, the response to price changes depends on the type of product. For example, increases in the price of staple foods eaten daily, such as maize, rice, roots and tubers or green bananas, have a relatively small effect on volumes of sales, as people still need to eat. Similarly, falls in price will not lead to major consumption increases. This is not the case with luxury or nonessential food items such as oranges and apples. Small falls in price can cause a disproportionately large increase in sales, while small price increases can lead to a big fall in sales.

Condiments, spices and other products that are used in small quantities are relatively insensitive to changes in price. Low prices do little to increase sales as people do not change their recipes to use more spices when prices are low. This means that when products like garlic are in oversupply, prices fall dramatically, as consumers do not respond to price changes by buying more.

Price of Changes in Demand

Consumer demand and tastes change constantly. Short-term changes can be caused by the weather. In northwest Europe, sunny days increase the demand for salad crops (tomatoes, lettuce, cucumbers) but when the weather becomes cooler vegetables for cooking are in stronger demand. This leads to short-term fluctuations in the prices of these commodities.

Longer-term changes in demand are caused by changes in taste, attitude and society. For example, perceived health benefits have led to long-term growth in the sales of broccoli and garlic (also considered to be good for the heart).

In Europe there has been a fall in the sale of fresh vegetables that need to be peeled and cooked and an increase in demand for semi-processed, cleaned, easier-to-prepare products such as frozen peas and carrots and frozen potato chips. This is because people are now reluctant to spend as much time cooking as they used to. Richer European consumers are often described as "cash rich-time poor".

While prices of oranges have fallen, those of easy-peeling types of citrus have increased because consumers prefer the convenience of mandarins and clementines.

Increasing Income on Sales

Food consumption patterns change as the amount of money that consumers have to spend increases.

— Consumption of cereals barely increases as incomes improve.

— Sales of milk products and beverages increase rapidly as incomes increase and, at higher income levels, these become the first and third most important expenditure categories.

— Fruit sales increase most rapidly with increased income.

Information such as this helps to indicate the relative importance of specific products and identifies which products are likely to increase in sales as economies develop. Farmers in countries with growing economies need to be aware of these opportunities and extension workers need to have the knowledge to help them take advantage of the changes that are taking place.

Marketing Task

Working with farmers, your main marketing task should be to provide them with the necessary information and contacts to allow them to increase their earnings through better marketing.

This involves helping them to:

— select profitable crops and improve production techniques;

— improve sales and achieve better prices;

— reduce costs and losses.

Crop Selection and Production

You need to know which crops generate better profits and how production choices, such as variety, production timing and product quality, can affect the profitability of individual crops. To do this you must know the products produced in your area and how they are marketed.

This information can be gathered by talking to farmers and traders individually or by meeting with them as a group. Individual interviews make it easier to gather confidential commercial information. They also make it easier for less confident farmers and those with lower status to express their views. On the other hand, gathering information from a group of farmers is effective and also very useful in helping farmers to understand their own problems and opportunities.

In the majority of cases your work will involve you in finding ways of improving the marketing of existing products. However, there may be the opportunity for farmers to diversify into new products and you also need to be in a position to identify market opportunities for these.

It is essential for farmers to know where and how they will sell their products before they start to grow them.

Improve Farmers' Sales

From speaking to farmers, you should be able to learn what they consider to be their main marketing problems and what ideas they have for solving those problems.

You can improve farmers' sales by working with them to improve their access to buyers and markets and by helping them to understand which products have expanding demand. You also need to know the expected prices for products and how prices change during the season. This will help farmers negotiate better prices by being better informed.

Improving sales and prices requires an understanding of the market. At a local level, this comes from talking to buyers, traders and transporters. Initially, you may be nervous about talking to business people. However, traders generally enjoy talking about their business and become very helpful if they feel this assistance will eventually lead to an increase in their turnover.

Carrying out market research in this way is a powerful tool for improving marketing. Once you start talking directly to the trade and understand how the market works, you can realise how valuable this information is to farmers and how much better equipped you are to help them.

With some support and help, farmers can also carry out their own market research. Once they have been helped to start finding out about the market for themselves they gain confidence and quickly feel able to do the work on their own. A farmer group can select between four and six farmers to carry out market research. You can advise them on interviewing techniques, provide them with checklists of questions in their own language and let them observe you doing market research interviews.

Information to Help Farmers

Farming Area

You need to find out the strengths and weaknesses of your area. This includes identifying:

— major crops grown, when they are harvested and their particular characteristics;

— costs of producing these crops;

— typical prices that farmers receive, how the prices change depending on buyer and time and what premiums for quality are given, if any;

— farmers' skills and their problems;

— businesses that transport, store, process and pack agricultural products;

— relative importance of the different local markets and traders;

— levels of existing trade and the possibility of growth;

— leaders in the farming community;

— assistance that farmers believe they require to help them market their products.

The purpose of gathering this information is for you to thoroughly familiarise yourself with both the problems and the opportunities of your area. It is very important to speak to farmers. In particular you must ensure that you meet medium- and small-scale farmers and not just large farmers. Farmers generally have a good understanding of their problems and are delighted to have an opportunity to talk. At this stage your role is to listen and learn. You should try to understand how they might react to new ideas and which farmers are likely to be most receptive to the idea of trying out new activities.

At the end of this information gathering you should have a clear idea of the crops, the marketing system, the individuals and the problems of the area. You should also have some idea of possible solutions that are worth investigating.

Market

Market research requires two main activities: firstly, finding out about how the market works and, secondly, finding out what products the market demands.

Finding out about how the market works involves understanding:

— the channels through which products pass;

— who the important people and companies are;

— how they do their business and who they currently buy from;

— their interest in trading with farmers from your area;

— who they sell to.

Finding out what products the market demands involves understanding:

— Product specifications, varieties, colour, size, grade, quality and packing;

— Prices, price patterns, variations according to season, quality and supply;
— Supply, volumes, competing suppliers and seasonality;
— Preferences of customers and consumers;
— Opportunities for additional production to be marketed.

Gathering Information about the Market

Interviews must be held with a cross-section of the farming community and with the small local businesses that are involved in agricultural marketing.

Gathering information about the market involves talking to the important businesses in the marketing chain, such as buyers who travel from farm to farm, transporters, other traders, wholesalers, processors, retailers, exporters and, less often, importers and even consumers. It should involve visiting local retail and assembly markets and, if applicable, major wholesale markets.

When interviewing traders and others it is essential to leave sensitive questions until you have the interviewee's confidence

Interviews should be carried out face-to-face but, when long distances are involved, telephone interviewing could be used if the people being interviewed are willing to cooperate over the phone.

To help gather information in a systematic way and to ensure that all the important questions are asked, it is useful to have a questionnaire or, at least, a checklist of questions. The following paragraphs briefly describe these questionnaires.

Farmer questionnaire. The first part covers questions to the farmer about the farm business and concentrates on gathering basic data needed. It asks such questions as how large the farm is, whether it has irrigation and what equipment the farmer has. It asks what major crops are grown, which ones are marketed and which are mainly for home consumption, Finally, it examines how the selling is normally done and asks what the farmer's major problems are.

The second part of the questionnaire concentrates on individual products, with questions on yields, costs, prices, packaging, transport and marketing arrangements. All of this is important information needed to build up knowledge about the crops and other agricultural products and to be able to calculate the costs and possible returns. It is not advisable to gather detailed production data on more than two products from each grower, as the process of data collecting can be very time consuming.

Trader questionnaire. The first part of the questionnaire covers questions to the trader about his or her business. The second part covers the most important agricultural products traded by that trader. It is not advisable to gather detailed data on more than two products per trader.

The most difficult part of carrying out an interview programme is the first interview. For example, you may feel nervous about talking directly to the private sector. It helps to have tested the questionnaire before starting the full interview programme. This will establish whether the wording of questions need to be changed and give you a chance to practice. Interviewing a small-scale farmer or a shopkeeper is a good way to start. It helps you to become comfortable with the process and to gain skills in interviewing.

When interviewing a group of farmers it is important to be well organised. Normally it is necessary to visit or send a message in advance, at least a day before, to arrange for the farmers' group to meet you at a specific place and time. You should speak to the farmers as a group to:

— explain the purpose of the survey;

— ask the farmers to outline the main crops that they grow and the problems they are experiencing.

When the meeting is finished you should thank the group and ask them if there are any additional points that they want to make, especially farmers who have not previously spoken.

When interviewing agribusinesses, traders and retailers the process is similar except that they are generally contacted and spoken to individually. This interviewing process involves:

— introducing yourself;

— explaining at the beginning why gathering the information is important (i.e. future plans must be based on a clear understanding of the market) and that this should be to the long-term advantage of the person being interviewed;

— explaining how much time it will take;

— starting by making the interviewees comfortable, often by encouraging them to explain how they started and developed their business;

— asking the noncontroversial questions first;

— keeping sensitive questions until towards the end of the interview;

— sharing information to encourage the interviewees to exchange their knowledge. For example, you could tell them about the production potential of farmers in your area;

— confirming the answer if there is any doubt as to what the person being interviewed means;
— writing down as much as possible on the questionnaire form while the interviewee is talking;
— thanking interviewees for their time, ideas and help;
— writing up the questionnaire properly and reading it to ensure your full understanding of what was said;
— interviewing a wide selection of farmers, traders and businessmen to be able to gather the ideas and opinions of a wide range of people.

Interview findings should be written up, in note form, as soon as possible. Use a table to compile the quantitative data. Let the data and the interview findings build up a picture of the farming sector and the rural resources. This should cover:

— the farm sizes and incomes;
— the products they produce;
— how farmers market their products;
— the important local agribusinesses supplying inputs and those involved in crop marketing;
— the differences in crops, markets and income between areas;
— typical costs of production and incomes for the main crops;
— farmers' problems, the solutions that they would like to see and the opportunities for developing the commercial aspects of their farming businesses.

For the major products that you ask questions about, you can learn:

— typical volumes and values of product sold;

— quality preference in terms of variety, appearance, size, taste, packing, grading;

— postharvest practices, shelf life, storage opportunities;

— differences in demand according to time of year, day of the week and whether influenced by festivals such as Diwali, Christmas, Lunar New Year or Ramadan;

— other competing suppliers, their quality, seasonality and price characteristics;

— trends in consumer demand and opportunities for additional supplies;

— typical prices and marketing costs.

As a result of conducting interviews you should have an understanding of the marketing chain and then be able to advise farmers on:

— the different marketing channels for major products and their relative importance;

— who the important businesses involved in agricultural marketing are and how/where they can be contacted;

— the different companies that growers can work with, how they like to do business with farmers and their trading and business interests;

— the likely costs of marketing, including transport, packaging, storage and commission charges;

— prices and how and why they change.

It is essential for farmers to know where and how they will sell their products before they start to grow them.

Message Across to Farmers

Talking directly to farmers, traders and others makes it

possible to gather information needed to improve agricultural marketing. Knowledge of the range of products produced, seasonality of crops and approximate costs of production can help farmers to understand their marketing possibilities and thus increase farm income. This first requires a careful analysis of the collected information and then an effective method of presentation of that information to farmers.

A crop calendar can be used to compare those periods when an area is able to supply produce with the periods when market prices are likely to be high. It can also be used to compare the seasonality of an area's production with that of a competing producing area. For example, farmers on the island of Mindanao in the southern Philippines have compared their crop calendar with that of competing vegetable farmers in Northern Luzon province, in order to identify periods when they have a production advantage.

Some crops, like cabbages, carrots and potatoes, which are temperate vegetables, are produced during the winter months. Melons and bananas, on the other hand, are mainly supplied during the hotter months.

Cost of production data and sales price figures can be used to draw up crop budgets. These can then be used to advise farmers on the potential profitability of different crops. Banks and credit institutions may also find the data useful in working out production loan requirements.

The yields that farmers claim to achieve can be very variable. Sometimes they don't know what their yields are. Yields on research stations are usually much higher than farmers' yields and must not be used for calculating crop budgets. Yields can be estimated by weighing the harvest from a known area of the field and then calculating the yield for a hectare. However, this is only useful when the research is being carried out at harvest time. Farmers can be asked to record yields in order to make future crop budget calculations possible.

Maps of production locations help extension workers and farmers to understand whether there are 'clusters' where particular products are produced. These maps are also useful for discussions with traders and can be used to identify locations where farmers could group together in order to sell to traders. Maps of alternative markets, distance from producers, size and whether their prices are good are also useful. They help producers to think about the options they have for selling their products.

In South Africa wholesalers are independent businesses who buy in bulk and sell on to the major retailers. Agents sell product on farmers' behalf in the wholesale (national produce) markets and receive a commission. Pick-up operators are small-scale mobile traders, driving small trucks, who buy both in the field and at wholesale markets and sell to hawkers, who sell fresh produce from stalls in small markets and on the streets.

In 2002, sales through both supermarkets and via the informal sector (i.e. pick-up traders and hawkers) were increasing, while sales through the wholesale markets were falling, although they remained the single most important outlet in terms of volume. Sales to the traditional processing sector (i.e. canning and drying) were flat but there was growth in more progressive types of processing such as snacks, ready-to-eat meals, sauces and convenience foods, as well as in exports.

One way of increasing the volume and value of transactions in the marketing chain is by improving the communication between the people in it. Farmers' requirements are often quite simple. They mainly want to know who would like to buy their products. Such information is not complex and lists of addresses and telephone numbers can be compiled with some short notes on each trader. Listing all the local traders, transporters and wholesalers of agricultural products can be very useful if farmers want to find alternative buyers for their products.

Farmers want to know what prices they could receive. Future prices are, of course, unknowable. For farmers, the best guide is the range of prices received in the past. Analysis of historical price data using averages and charts helps to indicate typical price patterns. Price data can be used to help farmers create simple crop budgets and to identify new market opportunities.

Diagnoses of marketing problems

Discussions with farmers can reveal their needs and problems, but first it is useful to study the local production and marketing practices. This makes it easier to understand what assistance farmers most need.

- in both areas vegetable production is the most important type of farming, but the cropping mix is slightly different;
- in Northern Province small farms are much better linked into the marketing chain, selling to hawkers and at open markets. In Eastern Free State farmers mainly sell to individual consumers;
- Northern Province producers hire transport or use their own, but Eastern Free State growers mainly rely on selling directly from their farms;
- Eastern Free State growers are generally nearer to

markets but complain about buyers demanding produce on credit and not paying, as well as the level of theft;

— Both areas believe that marketing could be improved by new physical markets and look forward to the expansion of supermarkets that would buy directly from farmers.

Table 1 Farmers' marketing constraints, Mozambique (percentage of farmers interviewed)

Constraints	District				
	Nampula	Mecuburi	Mogovolas	Ribaue	Malema
Transport	86	100	67	58	50
Lack of competition / few traders	33	66	78	66	19
Poor prices	24	50	45	33	25
Unreliable traders	29	33	45	33	19
Roads	76	33	-	-	25
Lack of, or poor market places	8	16	-	33	-
Farmer credit	-	33	-	-	25
Input costs	-	-	-	25	13

Table 1 summarises the perceived constraints of farmers in the northern part of Mozambique. This table was prepared following an interview programme with sixty farmers in five districts. Although the importance of the constraints identified varied from area to area, transport of produce was clearly considered to be the major one.

The second most important constraint was said to be the limited number of traders and the absence of competition between them. However, this raised the question of why there were so few traders. Was it because they did not have the necessary resources (i.e. cash, transport), knowledge and interest, or was it because there was little demand for the products the farmers grew.

Farmers always complain about poor prices and this is a predictable finding of any field research. Inadequate prices may reflect high marketing costs or a lack of demand. It may be possible to improve prices by lowering the unit costs of marketing and promoting more competition between buyers, by improving negotiating by farmers and by organising access to new market opportunities.

During the survey in Mozambique farmers often described traders as unreliable. By this they meant that the traders who visited were inconsistent and unpredictable and there was no certainty as to whether they would visit and what prices they would offer. This may indicate a wish on the part of farmers to develop more regular and reliable trade.

In addition to Trader Fact Sheets, Product Fact Sheets can be prepared to give specifications of the products demanded by the market, along with indications of the seasonal price and supply patterns. The purpose of the fact sheet is to provide producers and suppliers with the critical information they need in order to make correct decisions on the products they produce and sell.

The information presented in Product Fact Sheets can be developed during trader interviews. It should include information on:

— varieties, appearance, taste;

— grading and quality standards;

— seasonality and supply patterns;

— typical prices and price patterns;

— packaging;

— sales trends.

To Make Decisions

Extension workers sometimes try to "push" farmers into accepting recommendations. However, when decisions about what to grow or how to sell are imposed this rarely leads to success. Farmers do not feel that the decisions are theirs, they are not committed to them and they feel a lower sense of responsibility. If things do go wrong it is easy for farmers to blame the person who persuaded them to do the "wrong" thing.

Helping farmers to make their own decisions is a more difficult and slower process but, in the long run, it will be more successful than trying to tell farmers what to do. When groups of farmers take on "ownership" of their plans they are more enthusiastic, show more determination to overcome problems and take greater pride if their plan proves successful. They are much more able to overcome problems in the future and to actively seek solutions for themselves.

To make decisions, farmers must be well informed. They need to know what choices they have. They will have to discuss what they can do to improve their marketing and agree on what to do and on who is responsible for the individual tasks. Your role is to guide this process.

The ways in which you can help farmers to decide on what needs to be done to improve their incomes depends on what stage they have reached in their marketing development

Adapting Production

Your main role in marketing is to improve the understanding of your farmers about marketing and how they can become more commercial and profitable by producing crops that are demanded by the market. Most farmers are naturally conservative and will be reluctant to go into new enterprises, because they involve risk. New crops or the introduction of new technologies or production techniques should, therefore, initially be undertaken on a small-scale trial basis.

Growers need advice on which crops to grow and what market opportunities they can target. While the final decision must always be that of the farmers, you should be able to help them plan their production. Although production issues, such as labour availability and crop rotations, have to be taken into account, the key factor affecting production decisions is that production must be market-oriented. This means producing products for which there is a demand and which farmers can grow profitably.

Individual crop selection. Crop selection should be based on the likely net returns of the major alternative agricultural enterprises. These calculations will establish which products are likely to be the most profitable. They can be discussed with farmers.

Market research should have shown which local products are likely to be most successful in terms of cost, quality or seasonality, compared with products from other areas. The research should also have shown which varieties are favoured and the best time to supply them. These findings need to be converted into practical recommendations for farmers, covering such issues as:

— varieties favoured by the market;

— sowing dates (e.g. whether to extend the period of supply or aim for a particularly high-priced period). Whether to avoid times of oversupply;

— other techniques to extend production into high-priced periods, such as using late or early varieties, transplanting, or using polythene tunnels or irrigation;

— techniques to improve quality, such as optimum fertilization, crop protection, pruning, irrigation, and weather protection.

Table 2 sets out a summary of the comparative costs and returns of three different crops. Although Crop C returns the highest sales, its profitability is less than for

Crops A and B. Farmers should be aware of the difference between gross returns (i.e. the value of sales) and net returns (i.e. the value of sales after deduction of costs). They must consider likely net returns when planning what to produce.

Table 2 Selecting crops to grow

	Crop		
	A	B	C
Expected sales from 1 ha	3 000	2 000	4 000
Costs			
Production	1 000	500	2 000
Marketing	1 000	500	1 500
Net return	1 000	1 000	500
Select crop	Yes	Yes	No

In calculating costs it is useful to remember that 80 percent of costs are normally contributed by only 20 percent of the items. You should therefore concentrate on collecting accurate information on the major costs. Individual farmer costs and returns can vary greatly.

Selection of a range of crops. It is advisable for farmers to achieve a balance between growing a wide range of crops and concentrating on those products where they have the most advantage. Producing a range of crops reduces the impact of a possible production or market failure for one crop. For example, if farmers grow only potatoes and their crops are affected by Potato Blight, they will earn nothing. However, if they grow two or three other crops and have successful harvests they will be less affected by the potato disease.

Growing many different crops can cause problems. This is because the farmer is unlikely to have the necessary expertise in all the crops and the smaller volumes of each produced makes marketing more difficult. Very often, growers will have preferences for crops that they feel comfortable growing and/or that grow well on their land.

As agriculture becomes more developed, farming become more specialised. Individual farmers, such as those supplying supermarkets or agroprocessors, generally have to concentrate on relatively few crops. Although growers may become more skilled, they can rarely be experts in more than three or four crops.

The most profitable crops are often the most risky. It is useful to have a cropping system in which risky crops are balanced against crops that can be relied upon. An example might be to grow one or two large-volume vegetable crops, such as potatoes and onions, together with higher priced products such as capsicum (green or red peppers) and leafy vegetables or salads. Small farms often have more labour available per hectare than do large farms and they can take advantage of this by concentrating on growing labour-intensive crops. These are crops that cannot be harvested mechanically and may also require transplanting, pruning, hand weeding (hoeing) and harvesting on several occasions.

Farm Investments

Farmers are often tempted to make investments that are expensive but do not improve significantly the financial viability of their farm. This needs to be avoided. Decisions should have as priorities investments that can: assure an income, through reducing risk; increase incomes by improving prices; improve incomes by increasing yields; reduce production costs; or generate additional income.

Investments that can improve yield stability and hence reduce risks include:

— those that protect against pests and diseases, such as sprays and a sprayer;

— irrigation and other technology that reduces the adverse effects of the weather.

Such investments are particularly important in horticulture because when yields are low as a result of poor weather or pests, prices rise significantly. The grower whose yields are least affected makes the most profit.

Investments that improve prices include:

— technology for 'off-season' production, such as polythene tunnels and improved transplant technology;

— investments made directly in improving marketing, such as grading facilities, on-farm storage, pick-up trucks to transport produce and telephones to improve market communication;

— investments that improve the quality of produce, (e.g. irrigation, crop protection or high-quality planting material).

Investments that increase yields are:

— irrigation equipment;

— scientific fertiliser usage (e.g. correct quantities, optimum timing of application, right amounts of different nutrients);

— high quality planting material.

Investments that save on production costs include:

— herbicides and mechanical hoes, which can cut down on labour for weeding.

Investments that can create additional income are:

— processing equipment to add value to products;

— transport, to take products to new markets.

Input Supply

The supply of inputs can have a direct effect on profitability. Planting material is particularly important. Consumers can have strong preferences for particular varieties, colours and tastes. For example, in much of the Middle and Near East there is a preference for plum tomatoes. Product colour can also be important, e.g. red apples are sometimes preferred to green or golden varieties. Growers' returns can be improved by ensuring the supply of the correct planting material, and part of your role can be to discuss with nurseries and seed suppliers the varieties they should stock.

Pest and disease damage will seriously reduce a crop's price and its potential shelf life. Sometimes these problems can be solved by the correct crop protection practices. In India, spray programmes were introduced to control scab disease on apples in Kashmir. A crucial step in the successful introduction of this programme was ensuring that agricultural chemical shops had the recommended materials available.

Finance and Credit

Critical production constraints are often shortage of working capital and funds for investment purposes. Potential sources of funds, apart from the farmer's own resources, can be divided into two: formal and informal.

Formal sources. These are mainly banks and other types of financial institutions such as credit unions, savings and credit cooperatives and various types of microfinance organisations. They offer different types of loan products and normally apply market-based interest rates. In the case of banks, they require specific types of collateral in order to grant loans and credits, which, together with often lengthy loan procedures, tend to seriously limit most farmers' access.

Providing banks with cost-of-production budgets, detailed information on likely returns and other relevant information about the activity at the time of the loan request will enable them to assess more accurately the risk involved and may shorten the procedures. Credit unions and similar types of member-owned financial institutions are, by their nature, more open to farmers requiring smaller production-oriented loans. But, similar to microfinance institutions, they often face problems in providing large numbers of loans, due to limited funds.

Financial institutions have different rules as to what loan purposes they can cover. Some will only provide input loans whilst others may also offer loans to meet labour costs. In rare cases banks offer marketing loans to farmers. Such loans are needed to cover the cost of harvesting, transport, packaging and even storage of crops. The advantage of such loans is that growers are not restricted in who they sell to, which is the case when they borrow from traders.

Where growers are under contract to supply produce to an agribusiness (e.g. food processor or exporter), production loans based on the hypothetical value of the crop can be made. These are called hypothecated loans. Under a loan hypothecation scheme the bank, or the processor, advances a proportion of the expected income as production credit (say 50 percent), without seeking any additional security.

Loan recovery can be made by the agribusiness, which deducts the outstanding debt from the grower's income. Unfortunately, such arrangements sometimes break down when the farmer decides to sell the crop to someone else who offers a higher price, thus not honouring the initial agreement. In such situations you should discuss with farmers the benefits of long-term reliable outlets for their products, and point out that such benefits may be lost if they seek short-term price advantages in this way.

Informal credit. In many countries the availability of formal funds from financial institutions is very limited and farmers often have to depend on informal sources. Sources of informal credit include moneylenders, family members, friends, traders and input suppliers. Informal loans are often made on the basis of close family links or mutual trust and are free of time-consuming bureaucracy.

In some countries, the most important source of informal credit is often the trader. The role of traders as sources of loans is much misunderstood. Traders provide credit to farmers to secure future supply and, therefore, income. The true costs of such credit to farmers are difficult to determine. Common criticisms are that high interest rates are charged and that growers who have borrowed money are forced into selling their produce at low prices.

In some cases this is no doubt true and, in the case of poor prices, farmers cannot switch to another trader or wholesaler. However, the opposite can also apply as farmers, knowing the trader has to buy from them, pay less attention to produce quality. Traders do not normally lend money to farmers in order to exploit them. They lend money in order to ensure that farmers will produce enough of a crop to meet the demand. Loans by traders are recovered simply by deducting the money advanced from value of the sales.

Access to Transport

Without adequate access to transport, farmers are at a disadvantage. They are dependent on visiting buyers. With transport, growers have control over what market the product is transported to and are therefore potentially in a far stronger marketing position. Improved efficiency in transport, e.g. larger loads, quicker turnaround times and better utilisation of capacities, are all proven methods of lowering costs and opening new market opportunities.

You can have an important role in helping growers to gain access to transport. This could involve introducing farmers to transporters, planning a collection route and helping organise the initial services. Growers who do not sell to visiting traders usually have to transport produce to market in hired lorries or pick-ups. The grower generally has to either pay a fixed price for the hire of the truck, no matter how little is transported, or is charged by the box or sack. Both systems can be inefficient.

If the lorry is not fully loaded the costs per unit go up. Cost savings can be achieved by encouraging farmers to share transport. You could help them to assemble produce on a particular day of the week at specified collection points.

When transport is charged by the unit, transporters will generally overload lorries in order to maximise their income. By assembling produce at one point so as to guarantee full loads, a fixed price for the lorry can be negotiated and the growers can themselves ensure that the transport is not overloaded, and their produce is not damaged. It is difficult to resist the temptation to try to squeeze extra containers into a vehicle and this is often done by traders, as well as by farmers. However, the benefits in terms of reduced transport costs are usually outweighed by losses in terms of damaged produce.

Transporters often overload lorries to maximise their income. Even the best-packed produce can be damaged

Generally, the larger the individual load (i.e. the larger the truck used) the cheaper the unit cost of transport. For example, in Pakistan, an 8-tonne lorry travelling from northern Punjab to Karachi used to cost 4 000 rupees (i.e. 500 rupees per tonne of produce) while a 20-tonne articulated lorry cost 7 000 rupees for the same trip (i.e. 350 rupees per tonne).

Farmers who are able to make investments in transport mainly buy small pick-up trucks. The unit costs of transport are therefore higher than when hiring space in a larger lorry, but such pick-up trucks do offer farmers the possibility to:

— transport produce to the market immediately after harvest;

— supply transport services to neighbouring growers;

— take produce for sale to farmers' markets;

— make direct sales to retailers and caterers.

Bibliography

Adams, P. and L.C. Ho., "Differential effects of salinity and humidity on growth and Ca status of tomato and cucumber grown in hydroponic culture", *Acta Horticulturae,* 1995.

Adams, P. and R. Holder., "Effects of humidity, Ca and salinity on the accumulation of dry matter and Ca by the leaves and fruit of tomato (*Lycopersicon esculentum*)", *Journal of Horticultural Science,* 1992.

Bakker, J.C., "The effects of temperature on flowering, fruit set and fruit development of glasshouse sweet pepper (*Capsicum annuum L.*)", *Journal of Horticultural Science,* 1989.

Benoit, F. and CeustermansN., "Growing pepper on ecologically sound substrates", *Acta Horticulturae,* 1994.

Berkelmann, B., Wohanka .W. and Wolf G.A., "Characterization of the bacterial flora in circulating nutrient solutions of a hydroponic system with rockwool", *Acta Horticulturae,* 1994.

Brugger, M.F., Short .T.H. and BauerleW.L., "An evaluation of horizontal air flow in six commercial greenhouses", *American Society of Agricultural Engineers,* 1987.

Calpas, J. T., "First confirmed report of pepper mild mottle virus in Alberta", *Greenhouse Coverings,* 1998.

De Koning, A.N.M., "Development and growth of a commercially grown tomato crop", *Acta Horticulturae,* 1989.

Fierro, A., Tremblay .N. and Gosselin .A., "Supplemental carbon dioxide and light improved tomato and pepper seedling growth and yield", *HortScience,* 1994.

Gauthier, L. GX:, "A small talk-based platform for greenhouse environmental control. Part I. Modeling and managing the physical system", *Transactions of the ASAE,* 1992.

Gul, A. and Sevican .A. "Effect of growing media on glasshouse tomato yield and quality", *Acta Horticulturae,* 1992.

Hand, D.W., Wilson.J.W. and Hannah .M.A., "Light interception by a row crop of glasshouse peppers", *Journal of Horticultural Science,* 1993.

Jespersen, L. M. and Willumsen.J., "Production of compost in a heat composting plant and test of compost mixtures as a growing media for greenhouse cultures", *Acta Horticulturae,* 1993.

Khah, E.M. and Passam.H.C., "Flowering, fruit set and development of the fruit and seed of sweet pepper (*Capsicum annuum L.*) cultivated under conditions of high ambient temperature", *Journal of Horticultural Science,* 1992.

Lin, W.C. and Block.G.S., "Determination of water vapor in a small air sample by a non-dispersive infrared gas analyzer", *HortScience,* 1997.

Manrique, L.M. "Greenhouse crops: A review", *Journal of Plant Nutrition,* 1993.

McGrady, J.J., and Cotter. D.J. "Fresh conifer bark reduces root-knot nematode galling of greenhouse tomatoes", *HortScience,* 1989.

Nederhoff, E. M. and Vegter.J. G., "Canopy photosynthesis of tomato, cucumber and sweet pepper in greenhouses: measurements compared to models", *Annals of Botany,* 1994.

Papadopoulos, A.P., and Pararajasingham.S., "The influence of plant spacing on light interception and use in greenhouse tomato (*Lycopersicon esculentum* Mill.): A review", *Scientia Horticulturae,* 1997.

Pressman, E., Moshkovitch H., Rosenfeld .K., Shaked R., Gamliel B. and Aloni.B., "Influence of low night temperatures on sweet pepper flower quality and the effect of repeated pollinations, with viable pollen, on fruit setting", *Journal of Horticultural Science & Biotechnology,* 1998.

Schon, M.K. "Effects of foliar antitranspirant or calcium nitrate applications on yield and blossom-end rot occurrence in greenhouse-grown peppers", *Journal of Plant Nutrition,* 1993.

Tremblay, N. and Gosselin.A. "Effect of carbon dioxide enrichment and light", *HortTechnology,* 1998.

Wilson, J.W., Hand D.W. and Hannah.M.A., "Light interception and photosynthetic efficiency in some glasshouse crops", *Journal of Experimental Botany,* 1992.

Index